WASH Economics and Financing

In Focus – Special Book Series

WASH Economics and Financing

Editors
Britta Augsburg and Tristano Sainati

IWA PUBLISHING

Published by

IWA Publishing
Republic – Export Building, 1st Floor
2 Clove Crescent
London E14 2BE, UK
Telephone: +44 (0)20 7654 5500
Fax: +44 (0)20 7654 5555
Email: publications@iwap.co.uk
Web: www.iwapublishing.com

First published 2021

British Library Cataloguing in Publication Data
A CIP catalogue record for this book is available from the British Library

ISBN: 9781789062465 (paperback)

Contents

Editorial: WASH Economics and Financing: towards a better understanding of costs and benefits

INTRODUCTION

The case for investment in water, sanitation and hygiene (WASH) has been convincingly made. WASH is essential for protecting public health, is a human right and investing in it is compelling from a fiscal and economic point of view. While the estimated $114 billion per year of capital investments required to meet universal access to safely managed water and sanitation services by 2030 is often portrayed as a hefty price tag, current best estimates of benefit-to-cost ratios leave little doubt about its value.

What is less clear is how to allocate resources efficiently. There is limited evidence on the cost efficiency and cost effectiveness of various policy and implementation choices. The underlying drivers of demand for (new) technologies and solutions are, for example, poorly understood, as is beneficiaries' willingness to pay (WTP), leading to open questions about pricing policies and sustainable business models.

This is in contrast to other infrastructure sectors, such as energy and transport, where active literature on the economics and financing of services has been more helpful in defining national and international policy. Our objective with this Special Issue is to encourage the WASH sector to follow suit and start to effectively engage and research these issues.

At the heart of this Special Issue therefore, are papers which highlight some of the specificities, and challenges of conducting full economic evaluations of WASH interventions, provide a deeper understanding of potential solutions, and present new findings on costs and outcome measures, thereby contributing towards a fuller picture of WASH cost-effectiveness.

OVERVIEW OF THE SPECIAL ISSUE

We started this editorial by saying that the case for WASH investment has been convincingly made, which implies, in economic terms, that the benefits of doing so outweigh the costs. This is generally accepted, despite the fact that many benefits are yet to be quantified or even shown to be statistically significant. Two important challenges in doing so are addressed in this special issue: the assessment of complementarities between (components of) WASH investments (and other investments) and the quantification of intangible and difficult to measure benefits. Orgill-Meyer (2020) sheds light on the interaction between village-level latrine coverage and the availability of school latrines and Augsburg & Rodríguez-Lesmes (2020) provide empirical evidence on household-level economic and non-pecuniary benefits of sanitation.

The complexity and heterogeneity of WASH systems implies not only non-monetary benefits but also non-monetary costs, or negative consequence, which have to be added to a full economic cost assessment of WASH. Monetary cost assessment is therefore particularly complicated in WASH, and the WASH sector has some way to go before providing a clear and transparent cost assessment. Indeed, WASH is characterised by highly empirical and applied investigations of transactions, as opposed to a more theoretical approach that is typical of economics and management disciplines, e.g. institutional theory, transaction cost theory, agency theory. Sainati *et al.* (2020) highlight the lack of costing standards and of unambiguous terminology in the WASH sector compared to other infrastructure sectors and takes a step towards closing this gap by introducing a proposed standard cost metric for urban sanitation. One of the difficulties in this process is the required heterogeneity in WASH infrastructures, not only across WASH components, but also within, as highlighted by Grant *et al.* (2020) for water schemes in the context of rural Vietnam. The same study also shows that few of the assessed schemes are profitable based on a 20-year design life assumption and can only be sustained with subsidies. The difficulty of WASH costing and financing is a recurrent theme: Deal & Sabatini (2020)

doi: 10.2166/washdev.2020.002

assess actual and potential market penetration and market share of small-scale water utilities in rural Ghana showing suboptimal market penetration even if water is free, driven by the perceived attractiveness of alternative sources. Nagpal *et al.* (2020) similarly highlight that alternative water sources pose challenges for cost recovery through tariffs and Abramovsky *et al.* (2020) demonstrate that existing tariff structures for piped water consumption fall well short of recovering the costs of service provision in ten analysed countries. There is a lot to learn and understand about effective financing instruments that can lead to a sustained supply of safely managed WASH services. The choice and combination of instruments can improve efficiency and effectiveness in service provision by – among other things – improving monitoring systems and fostering innovation (or not), as discussed in Howard & White (2020).

Although Abramovsky *et al.* (2020) reveal that subsidies delivered via the water tariff largely fail to achieve the goal of improving the accessibility and affordability of piped water among the poor, other studies demonstrate that other modes of subsidy remain relevant and justifiable in the WASH sector. This is largely because of public benefits or externalities: Radin *et al.* (2020) draw attention to this through a comprehensive cost-benefit analysis which shows that community-led total sanitation (CLTS) combined with a well targeted subsidy would be more effective in the context of Ghana, than CLTS only. Similarly, Peletz *et al.* (2020) compare the financial structure of approaches to determine the feasibility of expanding safe faecal sludge services to low-income areas in urban Kenya and reveals that even the approach found to be most cost-effective requires subsidies to be sustained.

This provides an example of the importance of understanding willingness to pay and the corresponding financial incentives needed to achieve desirable outcomes for policy or programme design. Tidwell (2020) demonstrates the difficulty in eliciting this information, showing important variation in the level of WTP for urban sanitation services depending on the choice of demand assessment method chosen.

And this is where we come full circle: Who is willing to pay for WASH services, and how much, depends on who benefits and by how much, and whether these benefits are internalised and understood.

REFLECTIONS ON THE CURRENT STATE OF THE ART OF RESEARCH IN WASH ECONOMICS AND FINANCE

This special issue is grounded in the belief that the use of economic evidence is, and should be, instrumental in transparent and consistent policy-making, based on efficient spending decisions. The special issue provided an opportunity to review, analyse and evaluate the current status of WASH economics and its use.

It is clear that the economics and finance of WASH is particularly challenging. Two reasons stand out: For one, the generic 'problem' of WASH is that it encompasses three areas – water, sanitation and hygiene – that each individually necessitates the matching of demand with supply, ensuring safe services and technology, at the right location and the right time. The fact that individual WASH components are closely interlinked and now also more and more seen as a relevant factor in integrated, multisectoral approaches to achieving SDGs introduces further complexity. Second, the decision about who pays how much for what, when and how needs to be considered in view of WASH carrying key traits of public goods, where much of the benefit is public. In other words, public finance of WASH can be justified by its inherent externalities. Little, however, is understood about how much *society* is willing to pay for promotion of health, dignity and a cleaner environment through WASH, and whether and how this WTP can be moved. There is a need to shift the academic and policy debate beyond the focus on the front end of WASH, the interface with beneficiaries, which includes an understanding of their WTP, and increase the scope to the back end of WASH system, including supply chains, overall business models, instrumental markets for the design and development of underlying infrastructure, and governance and institutional implications derived from economic and financial transactions in WASH.

One important learning coming from the special issue is that this complexity is clearly not an insurmountable problem, but rather an academically interesting and policy relevant challenge, that researchers are beginning to tackle. In order to achieve improved outcomes for the poor and services that are efficiently delivered, we need at the same time more, better, and better consolidated knowledge on the selection and financing of WASH interventions.

There are strong individual researchers working on it from their respective disciplinary angles, but bringing the research cohesively together is an achievement yet to be made. For example, many academic disciplines put a lot of efforts into obtaining robust estimates of WASH benefits, particularly in terms of health but increasingly also covering non-health outcomes. The sector has seen very productive and fruitful cross-disciplinary research projects doing so. What the majority of such efforts seem to be lacking though, is an equally robust estimation of costs. Rather than striving for an approach that can yield the underlying cost function with proper confidence intervals, a pure accounting perspective is typically taken to assess costs, at best.

To date, decisive and consistent answers to questions relevant for improving the delivery of WASH services are not available. Barriers to adopting new products, technologies, and solutions need to be identified systematically, underlying factors that affect demand need to be estimated to inform pricing policy, shed light on information gaps and the role of credit. The variability of these factors needs to be understood in different geographical, institutional, cultural, and political contexts. There is further a need to systematise and integrate the current academic debate so that the results can be directly used by practitioners and activists.

We believe that this special edition from the Journal of Water, Sanitation and Hygiene for Development makes a timely and useful contribution to this debate and stimulates further work on both the theory and practice of WASH economics and financing. We look forward to reading many new and challenging contributions to this debate in the years to come.

Guest Editors
Britta Augsburg
Institute of Fiscal Studies, UK

Tristano Sainati
School of Civil Engineering, University of Leeds, UK

REFERENCES

Abramovsky, L., Andrés, L., Joseph, G., Rud, J. P., Sember, G. & Thibert, M. 2020 Unpacking piped water consumption subsidies: who benefits? new evidence from 10 countries. *Journal of Water, Sanitation and Hygiene for Development* **10** (4), 691–715.

Augsburg, B. & Rodríguez-Lesmes, P. 2020 Sanitation dynamics: toilet acquisition and its economic and social implications in rural and urban contexts. *Journal of Water, Sanitation and Hygiene for Development* **10** (4), 628–641.

Deal, P. T. & Sabatini, D. A. 2020 An assessment of penetration for pay-to-fetch water kiosks in rural Ghana using the Huff gravity model. *Journal of Water, Sanitation and Hygiene for Development* **10** (4), 670–679.

Grant, M., Foster, T., Van Dinh, D., Willetts, J. & Davis, G. 2020 Life-cycle costs approach for private piped water service delivery: a study in rural Viet Nam. *Journal of Water, Sanitation and Hygiene for Development* **10** (4), 659–669.

Howard, G. & White, Z. 2020 Does payment by results work? Lessons from a multi-country WASH programme. *Journal of Water, Sanitation and Hygiene for Development* **10** (4), 716–723.

Nagpal, T., Rawlings, H. & Balac, M. 2020 Understanding water demand and usage in Mandalay city, Myanmar as a basis for resetting tariffs. *Journal of Water, Sanitation and Hygiene for Development* **10** (4), 680–690.

Orgill-Meyer, J. 2020 Interaction of village and school latrines on educational outcomes in India. *Journal of Water, Sanitation and Hygiene for Development* **10** (4), 618–627.

Peletz, R., Feng, A., MacLeod, C., Vernon, D., Wang, T., Kones, J., Delaire, C., Haji, S. & Khush, R. 2020 Expanding safe fecal sludge management in Kisumu, Kenya: an experimental comparison of latrine pit-emptying services. *Journal of Water, Sanitation and Hygiene for Development* **10** (4), 744–755.

Radin, M., Wong, B., McManus, C., Sinha, S., Jeuland, M., Larbi, E., Tuffuor, B., Biscoff, N. K. & Whittington, D. 2020 Benefits and costs of rural sanitation interventions in Ghana. *Journal of Water, Sanitation and Hygiene for Development* **10** (4), 724–743.

Sainati, T., Zakaria, F., Locatelli, G., Andrew Sleigh, P. & Evans, B. 2020 Understanding the costs of urban sanitation: towards a standard costing model. *Journal of Water, Sanitation and Hygiene for Development* **10** (4), 642–658.

Tidwell, J. B. 2020 Users are willing to pay for sanitation, but not as much as they say: empirical results and methodological comparisons of willingness to pay for peri-urban sanitation in Lusaka, Zambia using contingent valuation, discrete choice experiments, and hedonic pricing. *Journal of Water, Sanitation and Hygiene for Development* **10** (4), 756–767.

Research Paper

Interaction of village and school latrines on educational outcomes in India

Jennifer Orgill-Meyer

ABSTRACT

Research has shown that improved sanitation environments can lead to better educational outcomes. Similarly, there is some evidence that school latrines may also improve school enrollment and attendance, particularly for girls. This paper explores the interaction effect of village-level latrine coverage and school latrines on child educational outcomes. The overall improved sanitation conditions from higher latrine coverage and the presence of school latrines might produce an additive effect and improve educational outcomes even further. Using multiple years of data from the Annual Status of Education Report (ASER) in India, this paper shows that there is no evidence of an additive effect. However, across multiple models, village-level latrine coverage is associated with lower school dropout rates and higher test scores, particularly among girls. School latrines do not appear to have a strong positive association with educational outcomes.

Key words | education, gender, India, sanitation

Jennifer Orgill-Meyer
Department of Government and Public Health Program,
Franklin and Marshall College,
Lancaster, PA 17603,
USA
E-mail: *jorgill@fandm.edu*

INTRODUCTION

As of 2018, 40% of the rural Indian population practiced open defecation (UNICEF 2018). Open defecation has large health, time, and dignity costs (Checkley *et al.* 2004; Cutler & Miller 2005; Wolf *et al.* 2014; Caruso *et al.* 2015; Dickinson *et al.* 2015; Augsburg & Rodríguez-Lesmes 2018). Recent research has also shown that improved sanitation increases human capital by improving education outcomes (Ortiz-Correa *et al.* 2016; Spears & Lamba 2016; Adukia 2017; Orgill-Meyer & Pattanayak 2020). Accounting for these improved human capital benefits is important to consider in sanitation interventions.

In a systematic review by Sclar *et al.* (2017), the authors propose a causal pathway for the effects of sanitation on educational outcomes, including attendance/absenteeism, enrollment, and test scores. In the review, both household/community sanitation and school sanitation are shown to have an effect on school absences. Access to improved sanitation at the household or community level has been shown to lower the risk of fecal exposure, which in turn lowers the rates of infection and illness. These lower rates of infection and illness both improve cognitive development and increase the ability to attend school. This increased ability to attend school may result in increased school enrollment if the health returns are substantial, decreased school absenteeism, and improved school performance as measured by test scores. Sclar *et al.* (2017) also theorize that access to improved school sanitation provides a more comfortable learning environment and thus increases the desire to attend school. In this paper, I investigate the interaction between community sanitation and school sanitation on a range of educational outcomes in India. Before discussing this interaction, I first provide an overview of the literatures investigating: (a) the link between household/community sanitation and educational outcomes and (b) the link between school sanitation and educational outcomes.

doi: 10.2166/washdev.2020.049

Household/community sanitation and educational outcomes

There is a strong literature establishing the impact of improved sanitation environments on reductions in infection and illness in children (Wolf *et al.* 2014; Freeman *et al.* 2017). Recent research has highlighted the effect of reduced sanitation-related infections and illnesses on cognitive development. Pinkerton *et al.* (2016) show that diarrhea is a significant predictor of delayed cognitive development among a number of different measures. Cumulative fecal exposure contributes to environmental enteropathy, which inhibits the absorptive capacity of the gut even when otherwise asymptomatic (Petri Naylor & Haque 2014; Crane Jones & Berkley 2015; Mbuya & Humphrey 2016). Reductions in enteric infections contributing to environmental enteropathy are also correlated with improvements in cognitive development (Petri *et al.* 2008; Jiang *et al.* 2014; Watanabe & Petri Jr 2016; Kosek 2017).

Beyond the mechanism of infections contributing directly to cognitive development, a growing body of literature investigates the impact of household/community sanitation improvements on other educational outcomes. Reductions in illnesses and infections as a result of improved sanitation environments may increase the ability of children to attend school, which could in turn contribute to enhanced cognitive development (Sclar *et al.* 2017). Research investigating community-level sanitation improvements on school attendance provides mixed results. A piped sewer intervention in Yemen demonstrated no effect on school attendance (Klasen *et al.* 2012), whereas an improved drainage intervention in Pakistan was associated with significant improvements to school attendance particularly for girls (Rauniyar Orbeta & Sugiyarto 2011). Notably, both of these interventions were coupled with community-wide water supply improvements, making it difficult to disentangle the sanitation improvements on school outcomes. Ortiz-Correa *et al.* (2016) find that sewerage systems in Brazil increase the total years of schooling attained. Research investigating household-level sanitation rather than community-wide sanitation finds that household sanitation is significantly correlated with reduced school absences (Dreibelbis *et al.* 2013; Park *et al.* 2015) and improved school attendance (Battiston *et al.* 2013), though all of these studies are cross-sectional making causal claims difficult.

School sanitation and educational outcomes

School sanitation conditions also play a role in impacting educational outcomes. Access to latrines or toilets in school creates a more comfortable learning environment, thus enhancing the desire to attend school (Sclar *et al.* 2017). There is mixed empirical evidence on school sanitation and educational outcomes. Grant Lloyd & Mensch (2013) do not find that the presence of school latrines or latrine quality correlates with absenteeism in menstruating girls in Mali, and Caruso *et al.* (2014) find no impact of a latrine cleaning intervention on school absences in Kenya. School WASH intervention studies have been shown to decrease absenteeism (Freeman *et al.* 2011; Trinies *et al.* 2016), though it is difficult to assess the role of sanitation alone in these interventions. Dreibelbis *et al.* (2013) find that school latrine maintenance is correlated with reduced school absences for both boys and girls. Adukia (2017) and Garn *et al.* (2013) use school enrollment rather than absenteeism as their outcome of interest. Adukia (2017) finds that the presence of sex-specific latrines increased school enrollment and test scores of pubescent-age girls in India. One explanation for why school latrines may have an effect on school enrollment or attendance for pubescent-age girls is the additional privacy they offer to girls who are menstruating (Alam *et al.* 2017). Similarly, a school-based WASH intervention increased school enrollment particularly among girls in Kenya (Garn *et al.* 2013).

Interaction of community and school latrines on education outcomes

In this paper, I explore the interaction of the village sanitation environment and the school sanitation environment on educational outcomes in India. Most of the research to-date that explores the effect of improved sanitation on educational outcomes either uses sanitation improvements at the household/community level or school toilets/latrines as their variable of interest as discussed above and summarized in B1–B3 in Table 1. Only one paper investigates both school and household-level sanitation (Dreibelbis *et al.* 2013); however, they do not explore the interaction effect and rely on cross-sectional data. This paper adds to this existing literature by being the first to use a panel dataset

Table 1 | Predicted effects of the interaction of school and village latrines on education outcomes

	No school latrines	Presence of school latrines
Low village-level latrine coverage	B1: **Poor educational outcomes**	B2: More desirable learning environment → **Improved educational outcomes**
High village-level latrine coverage	B3: Decreased fecal exposure → decreased infection and illness → **Improved educational outcomes**	B4: **Hypothesized additive effect (>B2 + B3)**

Educational outcomes include: absenteeism/attendance, school enrollment, and test scores.

to investigate the interaction effect of village-level sanitation and school sanitation on educational outcomes. I use village-level sanitation rather than household sanitation because of the positive externalities that sanitation provides (Alderman Hentschel & Sabates 2003; Watson 2006; Fuller *et al.* 2016). Existing literature suggests that the primary mechanism through which sanitation improves educational outcomes is through reduced fecal exposure and thus reduced infections and illnesses (Pinkerton *et al.* 2016; Watanabe & Petri Jr 2016; Kosek 2017; Sclar *et al.* 2017). It is unlikely that reductions in fecal exposure from increased latrine adoption accrue to only the household that adopts the latrine. Using village-level sanitation rather than household sanitation accounts for the health externalities of an improved sanitation environment (Alderman Hentschel & Sabates 2003; Buttenheim 2008; Fuller *et al.* 2016; Geruso & Spears 2018).

As discussed above, there is a growing literature suggesting that both household/community sanitation and school sanitation are important predictors of educational outcomes as measured by attendance/absenteeism, school enrollment, and test scores, though some research shows weak or no effects. In this paper, the main hypothesis is that the interaction of improved village-level sanitation and school sanitation produces an additive effect on educational outcomes. In other words, does improved village-level sanitation and school sanitation produce educational outcomes over and above what one would observe with just improved village-level sanitation or school sanitation? I hypothesize that the more desirable learning conditions from school latrines combined with the reduced infection rates from village latrine coverage produce even greater educational returns (see Table 1). Using absenteeism as an example outcome, existing research shows that improved sanitation environments and school latrines are correlated with lower rates of absenteeism (Freeman *et al.* 2011; Rauniyar Orbeta & Sugiyarto 2011; Battiston *et al.* 2013; Dreibelbis *et al.* 2013; Park *et al.* 2015; Trinies *et al.* 2016). If a school latrine offers an additional incentive to attend school, the health conditions resulting from the improved sanitation environment may reduce absenteeism even further. The educational outcome variables that I use to test this hypothesis are school dropout rates and math test scores. I explain this choice of variables further in the following section.

There are two alternatives to the additive hypothesis. First, there may be no interaction effect – that is school latrines, and village sanitation environments produce improved educational outcomes independently and thus there is no complementary effect. Second, there may be a substitution effect, where, in the presence of both school latrines and improved village sanitation environments, educational outcomes actually worsen. The substitution effect would occur if the increased likelihood of having a latrine at home weakens the additional incentive provided by school latrines to enroll in or attend school. In this study, I make use of a large national Indian dataset to test the proposed additive effects shown in B4 in Table 1.

METHODS

Data overview

To investigate this research question, I use data from India's Annual Status of Education Report (ASER) survey. ASER is an annual survey, which collects data on childhood educational attainment in rural India. Specifically, ASER measures educational outcomes (school enrollment and educational attainment tests) for approximately 600,000 children ages 3–16. ASER surveys 30 villages per rural district and interviews 20 households per village in each year of data collection. Additionally, ASER collects data on one government school per village, gathering data on teaching indicators and school infrastructure (including school latrine status). Each year ASER randomly replaces 10 villages from each district. Pooling different years of ASER data results in a repeated cross-section at the household level and an unbalanced panel at the village and school level since most villages and schools are observed for multiple years.

Statistical model

The model used to test the effect of the interaction of school latrines and village latrine coverage on education outcomes is estimated with the ordinary least squares model shown below.

$$\begin{aligned}\text{Outcome}_{ijt} = {} & \beta_0 + \beta_1 \text{Vil\_Latrine}_{jt} + \beta_2 \text{School\_Latrine}_{jt} \\ & + \beta_3 \text{Female}_i + \beta_4 \text{Vil\_Latrine}_{jt} \times \text{School\_Latrine}_{jt} \\ & + \beta_5 \text{Vil\_Latrine}_{jt} \times \text{Female}_i \\ & + \beta_6 \text{School\_Latrine}_{jt} \times \text{Female}_i \\ & + \beta_7 \text{Vil\_Latrine}_{jt} \times \text{School\_Latrine}_{jt} \times \text{Female}_i \\ & + \beta_8 \text{Age}_{ijt} + \gamma \text{HHchars}_i + \theta_j + \delta_{kt} + \varepsilon_{ijt}\end{aligned}$$

I use two measures of educational outcomes as the dependent variables in the above model: (1) math test scores and (2) a binary variable for whether the child has dropped out of school. ASER does not measure reported school attendance or absenteeism – the two measures often used in the literature on WASH and educational outcomes (Rauniyar Orbeta & Sugiyarto 2011; Battiston *et al.* 2013; Dreibelbis *et al.* 2013; Caruso *et al.* 2015; Park *et al.* 2015; Trinies *et al.* 2016). I use math test scores as a proxy for attendance and absenteeism with the assumption that children that attend school more frequently will score higher on these tests. The score, as measured by ASER, ranges from 0 to 4, with 0 meaning that the child was unable to complete any of the questions accurately on the test, 1 indicating that the child correctly identified a number between 1 and 9, 2 representing that the child correctly identified a number between 10 and 99, 3 representing that the child correctly solved a subtraction problem, and 4 indicating that the child correctly solved a division problem. Math test scores have been used as a proxy for impact of sanitation on educational attainment when absenteeism and attendance metrics are unavailable (Spears & Lamba 2016; Adukia 2017). The school dropout indicator acts as a proxy for school enrollment since children who have dropped out of school will not be enrolled. While there is some evidence of the impact of school latrines on female enrollment (Garn *et al.* 2013; Adukia 2017), there is no literature that studies the impact of community sanitation environments on school enrollment. Ortiz-correa *et al.* (2016) do find that community-wide sanitation improvements increase overall years of completed schooling. I use a linear probability model for this outcome variable, though results are similar when using probit and logistic regressions.

Vil_Latrine_{jt} measures the proportion of households in village j in year t that have latrines. Since ASER randomly samples 20 households from each village included in the survey, this variable should approximate actual village-level latrine coverage. $\text{School\_Latrine}_{jt}$ is an indicator variable with one representing whether there is a usable girls' toilet or latrine present at the school in village j at time t. Since the literature on school latrines and educational outcomes has found that separate-sex latrines are important to create enrollment or attendance incentives, particularly for girls (Garn *et al.* 2013; Adukia 2017), usable girls' toilets or latrines are the relevant variable for this analysis. Female_i is an indicator variable taking the value of one if the child is a female and zero if the child is a male. Interactions between these three independent variables are represented by β_4, β_5, β_6, and β_7. The coefficients of interest are β_4 which measures the effect of the interaction of school latrines and village latrine coverage on the outcome variables, and β_7 which measures whether this interaction effect differs by gender.

Using the model from Equation (1), I include the full set of school-age children in the ASER data (ages 6–16). Included as controls are the child's age, Age_{ijt}, and a vector of household characteristics, HHchars_i, specifically whether the child's mother attended secondary school (Black Devereux & Salvanes 2005; Carneiro Meghir & Parey 2013) and an index of household wealth discussed below. θ_j represents village fixed-effects which control for time-invariant village characteristics and δ_{kt} represents state (k) by year fixed-effects which control for state-level time trends. The ability to include village fixed-effects is an important feature of ASER data, which allows for the control of many village-level confounding factors that may also affect educational attainment.

The ASER household survey includes a series of asset questions measuring household wealth. Rather than including multiple asset ownership indicator variables as a control, I used principal component analysis (PCA) to create an asset index for each household. The assets included in each index are: household latrines or toilets, electricity connection, televisions, and mobile phones. Using a PCA-created asset index is more appropriate than including separate asset variables when assets are correlated, which they are in the ASER data (Vyas & Kumaranayake 2006). Controlling for household sanitation is important to isolate the main effect of village-wide sanitation (Alderman Hentschel & Sabates 2003; Fuller *et al.* 2016); however, we also conduct a sensitivity analysis omitting household sanitation from the asset index and do not find significant differences in our main result (see Supplementary Tables 1A and 2A).

Data overview

I use annual ASER data from 2012, 2013, 2014, and 2016 (ASER data were not collected in 2015); I do not use earlier rounds of ASER data because information on school latrines was not collected prior to 2009 and the measurement for school dropout rates is inconsistent prior to 2012. As a result, I focus on the time period 2012–2016 in my analysis. For the math score outcome, I drop 2012 in the analysis since the metrics that ASER used to measure math scores changed in 2013. I drop any child observations in villages where data on usable female latrines in schools were missing, leaving a total sample size of 2,774,585 observations. Mother's education was missing for certain households, so the total sample size drops to 2,597,011 observations when including those control variables.

As shown in Table 2, both village-level latrine coverage and the percentage of schools with usable girls' toilets/latrines have increased dramatically over time between 2012 and 2016. These trends are also reflected in Figure 1 (which also shows pre-trends starting in 2009). The school dropout rate, on the other hand, does not appear to have changed over time. While the dropout rate is low (3.4% in 2016), ASER only measures if a student is officially not enrolled in school, and thus, this dropout rate does not capture low or no attendance among enrolled students.

Table 2 | Summary statistics over time

Variable	% in 2012 (n = 392,440)	% in 2016 (n = 400,954)
Village latrine coverage	40.3	53.5
Schools with usable girls' toilets/latrines	47.2	60.4
Children (ages 6 and above) having dropped out of school	3.4	3.4
Child is female	47.3	48.4
Mother attended secondary school	50.2	54.4
Household has electricity connection	75.6	82.5
Household has TV	48.8	54.2
Household has mobile phone	67.4	80.2

RESULTS AND DISCUSSION

Table 3 contains the main results for the school dropout outcome variable from the different models of Equation (1) outlined in the previous section. In each of the models, village latrine coverage is associated with a lower school dropout rate. According to Model 1, moving from no village latrine coverage to full coverage is associated with a 3.3% school dropout rate. The interaction between *Village latrine coverage* and *Female* also indicates that this effect is stronger for female children. This finding is consistent with other literature, suggesting that educational impacts of latrine coverage are more concentrated among female children (Garn *et al.* 2013; Adukia 2017; Orgill-Meyer & Pattanayak 2020).

Having a usable girls' latrine in the school is only associated with a statistically significant reduction in school dropout rates in two models (columns 2 and 4). Moreover, the effect does not appear to differ between female and male children. This finding suggests that overall village latrine coverage matters more for improving school enrollment than the presence of school latrines. The main result, the additive effect of village latrine coverage and school latrines, represented by the interaction *Village*

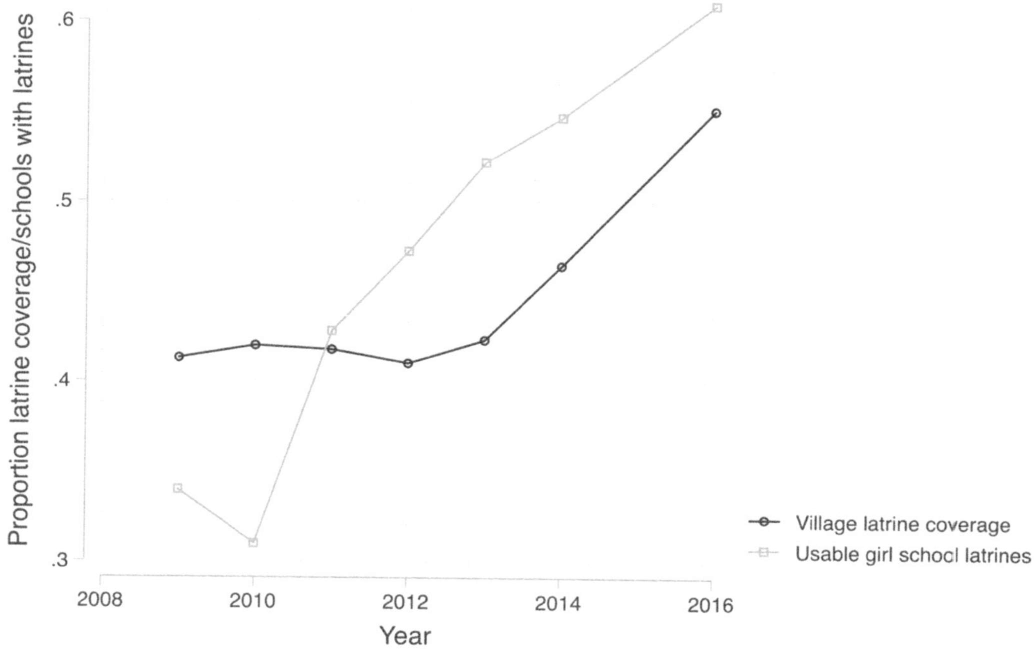

Figure 1 | Trends in latrine coverage (2009–2016).

Table 3 | Effect of latrine coverage and school latrines on school dropout rates

	(1)	(2)	(3)	(4)
Village latrine coverage	−0.033*** (0.001)	−0.044*** (0.004)	−0.010*** (0.004)	−0.009** (0.004)
School latrine	0.000 (0.001)	−0.006* (0.003)	−0.001 (0.003)	−0.005* (0.003)
Female	0.018*** (0.002)	0.010*** (0.002)	0.010*** (0.002)	0.011*** (0.002)
Village latrine coverage × female	−0.014*** (0.002)	−0.005** (0.002)	−0.004** (0.002)	−0.005** (0.002)
School latrine × female	−0.004 (0.003)	−0.000 (0.003)	−0.000 (0.003)	−0.001 (0.003)
Village latrine coverage × school latrine	−0.005*** (0.001)	0.003 (0.006)	−0.000 (0.005)	0.006 (0.005)
Village latrine coverage × school latrine × female	0.004 (0.003)	0.001 (0.003)	0.001 (0.003)	0.001 (0.003)
Child age	0.014*** (0.000)	0.014*** (0.000)	0.013*** (0.000)	0.013*** (0.000)
Mother attended secondary school			−0.021*** (0.000)	−0.021*** (0.000)
Household asset index			−0.014*** (0.000)	−0.014*** (0.000)
Constant	−0.096*** (0.001)	−0.093*** (0.002)	−0.099*** (0.002)	−0.095*** (0.010)
Observations	1,549,430	1,549,430	1,448,859	1,448,859
R^2	0.052	0.127	0.135	0.135
Village FE	No	Yes	Yes	Yes
State × year FE	No	No	No	Yes

Robust standard errors in parentheses; standard errors clustered at the village level. School latrine is defined as having a usable female latrine in the school ***$p < 0.01$, **$p < 0.05$, *$p < 0.1$.

Latrine Coverage × *School Latrines*, is negative and statistically significant in the base model (column 1). However, the coefficient does not have statistical significance when including village fixed-effects and other controls (columns 2–4), suggesting no evidence of an additive effect. There are also no differences in this interaction by gender.

Conducting the same analysis for each state in India reveals positive additive effects for only one Indian state: Kerala. This lack of additive effect at the state level further suggests that village latrine coverage and school latrines have independent channels of producing educational benefits. When looking at just the village latrine coverage effects on school dropout rate, there is a lot of heterogeneity at the state level as depicted in Table 4. This table contains the coefficients on village latrine coverage for each state. Most of the literature on village/household sanitation has used absenteeism and attendance as the dependent variables (Rauniyar Orbeta & Sugiyarto 2011; Battiston *et al.* 2013; Dreibelbis *et al.* 2013; Park *et al.* 2015). This finding is the first that suggests that village-level latrine coverage also can have an impact on school dropout rates. For the most part, the larger effect sizes are concentrated in the relatively wealthier states (with some exceptions – Bihar has a small but significant effect size), suggesting a possible threshold effect.

Table 5 depicts the same models with math scores as the dependent variable. Recall that math scores are being used as a proxy for attendance/absenteeism which is not captured in the ASER data. In the base model (column 1), both village latrine coverage and school latrines are associated with higher math scores, though these coefficients lose significance when including fixed-effects and other controls. Similar to the school dropout outcome, the effects of having higher village latrine coverage on math scores are higher among female children (consistent across all four models). However, there are no differential impacts of having school latrines by gender. Finally, there does not appear to be a statistically significant effect on the interaction of village latrine coverage and school latrines; nor do these coefficients differ by gender.

Taken together, there does not appear to be strong evidence of either an additive effect of village-level latrine coverage and school latrine on educational outcomes for either girls or boys. This lack of significance in the interactions between school and village latrines suggests that the channels through which village latrine coverage and school latrines affect educational outcomes are independent. While there does not appear to be support for an additive effect among school latrines and village latrine coverage, village-level latrine coverage does appear to consistently be associated with improved educational outcomes – reduced dropout rates and higher math scores – particularly among female children. This finding is significant when considering the economic gains that female education produces (Knowles *et al.* 2002); an important policy lever to achieving such gains could be to improve the overall sanitation environment.

Table 4 | Village-Level Latrine Coverage Effects on School Drop Out Rates by State

State	Coefficient	(Std Error)
Andhra Pradesh	-0.018**	(0.007)
Arunachal Pradesh	-0.006	(0.008)
Assam	0.010*	(0.005)
Bihar	-0.008*	(0.004)
Chhattisgarh	0.014*	(0.008)
Dadra & Nagar Haveli	-0.044	(0.035)
Daman & Diu	-0.009	(0.011)
Goa	-0.007	(0.007)
Gujarat	-0.010	(0.008)
Haryana	-0.027**	(0.011)
Himachal Pradesh	0.013	(0.011)
Jammu and Kashmir	-0.010**	(0.005)
Jharkhand	-0.001	(0.008)
Karnataka	-0.017***	(0.004)
Kerala	-0.134	(0.088)
Madhya Pradesh	-0.001	(0.005)
Maharashtra	-0.016***	(0.004)
Manipur	-0.001	(0.009)
Meghalaya	-0.019*	(0.011)
Mizoram	-0.009	(0.008)
Nagaland	0.014	(0.010)
Odisha	-0.012	0.008)
Pondicherry	0.020	(0.017)
Punjab	-0.004	(0.011)
Rajasthan	-0.019***	(0.007)
Sikkim	-0.052	(0.069)
Tamil Nadu	-0.006*	(0.003)
Telangana	-0.033	(0.023)
Tripura	-0.027	(0.022)
Uttar Pradesh	0.032***	(0.004)
Uttarakhand	0.014	(0.011)
West Bengal	0.011	(0.008)

*$p < 0.01$, **$p < 0.05$, ***$p < 0.001$

Table 5 | Effect of latrine coverage and school latrines on math scores

	(1)	(2)	(3)	(4)
Village latrine coverage	0.503*** (0.018)	0.377* (0.225)	0.243 (0.222)	−0.006 (0.269)
School latrine	0.061*** (0.013)	0.117 (0.184)	0.094 (0.176)	0.008 (0.176)
Female	−0.153*** (0.019)	−0.129*** (0.015)	−0.134*** (0.016)	−0.133*** (0.016)
Village latrine coverage × female	0.081*** (0.021)	0.058*** (0.016)	0.059*** (0.017)	0.059*** (0.017)
School latrine × female	−0.003 (0.028)	−0.012 (0.023)	−0.006 (0.024)	−0.006 (0.024)
Village latrine coverage × school latrine	0.009 (0.024)	0.023 (0.334)	−0.062 (0.321)	0.065 (0.319)
Village latrine coverage × school latrine × female	0.022 (0.030)	0.030 (0.024)	0.024 (0.025)	0.025 (0.025)
Child age	0.262*** (0.001)	0.259*** (0.001)	0.261*** (0.001)	0.261*** (0.001)
Mother attended secondary school			0.228*** (0.005)	0.227*** (0.005)
Household asset index			0.163*** (0.004)	0.162*** (0.004)
Constant	−0.716*** (0.012)	−0.678*** (0.101)	−0.641*** (0.098)	−0.425*** (0.126)
Observations	329,689	329,689	310,038	310,038
R^2	0.309	0.534	0.550	0.550
Village FE	No	Yes	Yes	Yes
State × year FE	No	No	No	Yes

Robust standard errors in parentheses; standard errors clustered at the village level. School latrine is defined as having a usable female latrine in the school. $***p < 0.01$, $**p < 0.05$, $*p < 0.1$.

This study also faces a number of limitations. The changes in how ASER data measured school dropout and math scores restrict the number of years in the analysis. Further, the sampling strategy used by ASER produces an unbalanced panel at the village level and repeated cross-sections at the household level. As a result, the analysis can utilize village fixed-effects which control for time-invariant level factors but cannot use household fixed-effects. Controlling for maternal education and household wealth, two factors that contribute to child educational attainment, allows for the elimination of some omitted variable bias, but the analysis still does not produce fully causal results.

CONCLUSIONS

This paper uses a large dataset from India to analyze the effect of the interaction of village-level latrine coverage and school-based latrines on child educational outcomes. There does not appear to be strong evidence for an additive effect of village and school latrines on school dropout rates or math test scores. All models explored in this paper produce positive associations between village-level latrine coverage and educational outcomes – lower dropout rates and higher test scores, particularly for female children. The relationship between school latrines and educational outcomes appears weaker.

ACKNOWLEDGEMENTS

Gaone Moetse provided valuable research assistance in dataset creation.

SUPPLEMENTARY MATERIAL

The Supplementary Material for this paper is available online at https://dx.doi.org/10.2166/washdev.2020.049.

REFERENCES

Adukia, A. 2017 Sanitation and education. *American Economic Journal: Applied Economics* **9** (2), 23–59.

Alam, M. U., Luby, S. P., Halder, A. K., Islam, K., Opel, A., Shoab, A. K., Ghosh, P. K., Rahman, M., Mahon, T. & Unicomb, L. 2017 Menstrual hygiene management among Bangladeshi adolescent schoolgirls and risk factors affecting school absence: results from a cross-sectional survey. *BMJ Open* **7** (7), 1–10.

Alderman, H., Hentschel, J. & Sabates, R. 2003 With the help of one's neighbors: externalities in the production of nutrition in Peru. *Social Science & Medicine* **56**, 2019–2031.

Augsburg, B. & Rodríguez-Lesmes, P. A. 2018 Sanitation and child health in India. *World Development* **107**, 22–39.

Battiston, D., Cruces, G., Lopez-Calva, L. F., Luga, M. A. & Santos, M. E. 2013 Income and beyond: multidimensional poverty in six. *Social Indicators Research* **112** (2), 291–314.

Black, S. E., Devereux, P. J. & Salvanes, K. G. 2005 Why the apple doesn't fall far: understanding intergenerational transmission of human capital. *American Economic Review* **95** (1), 437–449.

Buttenheim, A. M. 2008 The sanitation environment in urban slums: implications for child health. *Population and Environment* **30** (1–2), 26–47.

Carneiro, P., Meghir, C. & Parey, M. 2013 Maternal education, home environments, and the development of children and adolescents. *Journal of the European Economic Association* **11**, 123–160.

Caruso, B. A., Freeman, M. C., Garn, J. V., Dreibelbis, R., Saboori, S., Muga, R. & Rheinegans, R. 2014 Assessing the impact of a school-based latrine cleaning and handwashing program on pupil absence in Nyanza Province, Kenya: a cluster-randomized trial. *Tropical Medicine and International Health* **19** (10), 1185–1197.

Caruso, B. A., Sevilimedu, V., Fung, I., Patkar, A. & Baker, K. K. 2015 Gender disparities in water, sanitation, and global health. *The Lancet* **386**, 650–651.

Checkley, W., Gilman, R. H., Black, R. E., Epstein, L. D., Cabrera, L., Sterling, C. R. & Moulton, L. H. 2004 Effect of water and sanitation on childhood health in a poor Peruvian peri-urban community. *Lancet* **363**, 112–118.

Crane, R. J., Jones, K. D. J. & Berkley, J. A. 2015 Environmental enteric dysfunction: an overview. *Food and Nutrition Bulletin* **36** (10), S76–S87.

Cutler, D. M. & Miller, G. 2005 The role of public health improvements in health advances: the twentieth-century United States. *Demography* **42** (1), 1–22.

Dickinson, K. L., Patil, S. R., Pattanayak, S. K., Poulous, C. & Yang, J. H. 2015 Nature's call: impacts of sanitation choices in Orissa, India. *Economic Development and Cultural Change* **64** (1), 1–29.

Dreibelbis, R., Greene, L. E., Freeman, M. C., Saboori, S., Chase, R. P. & Rheingans, R. 2013 Water, sanitation, and primary school attendance: a multi-level assessment of determinants of household-reported absence in Kenya. *International Journal of Educational Development* **33** (5), 457–465.

Freeman, M. C., Greene, L. E., Dreibelbis, R., Saboori, S., Muga, R., Brumback, B. & Rheingans, R. 2011 Assessing the impact of a school-based water treatment, hygiene and sanitation programme on pupil absence in Nyanza Province, Kenya: a cluster-randomized trial. *Tropical Medicine & International Health* **17** (3), 380–391.

Freeman, M. C., Garn, J. V., Sclar, G. D., Boisson, S., Medilcott, K., Alexander, K. T., Penakalapati, G., Anderson, D., Mahtani, A. G., Grimes, J. E. T., Rehfuess, E. A. & Clasen, T. F. 2017 The impact of sanitation on infectious disease and nutritional status: a systematic review and meta-analysis. *International Journal of Hygiene and Environmental Health* **220**, 928–949.

Fuller, J. A., Villamor, E., Cevallos, W., Trostle, J. & Eisenberg, J. 2016 I get height with a little help from my friends: herd protection from sanitation on child growth in rural Ecuador. *International Journal of Epidemiology* **45** (2), 460–469.

Garn, J. V., Greene, L. E., Dreibelbis, R., Saboori, S., Rheingans, R. D. & Freeman, M. C. 2013 A cluster-randomized trial assessing the impact of school water, sanitation and hygiene improvements on pupil enrolment and gender parity in enrolment. *Journal of Water Sanitation and Hygiene for Development* **3** (4), 592–601.

Geruso, M. & Spears, D. 2018 Neighborhood sanitation and infant mortality. *American Economic Journal: Applied Economics* **10** (2), 125–162. https://doi.org/10.1257/app.20150431.

Grant, M., Lloyd, C. & Mensch, B. 2013 Menstruation and school absenteeism: evidence from rural Malawi. *Comparative Education Review* **57** (2), 260–284. https://www.journals.uchicago.edu/doi/abs/10.1086/669121.

Jiang, N. M., Tofail, F., Moonah, S. N., Scharf, R. J., Taniuchi, M., Ma, J. Z., Hamadani, J. D., Gurley, E. S., Houpt, E. R., Azziz-Baumgartner, E., Haque, R. & Petri Jr., W. A. 2014 Febrile illness and pro-inflammatory cytokines are associated with lower neurodevelopmental scores in Bangladeshi infants living in poverty. *BMC Pediatrics* **14** (50).

Klasen, S., Lechtenfeld, T., Meier, K. & Rieckmann, J. 2012 Benefits trickling away: the health impact of extending access to piped water and sanitation in urban Yemen. *Journal of Development Effectiveness* **4** (4), 537–565.

Knowles, S., Lorgelly, P. K. & Dorian Owen, P. 2002 Are educationl gender gaps a brake on economic development? Some cross-country empirical evidence. *Oxford Economic Papers* **54**, 118–149.

Kosek, M. N. 2017 Causal pathways from enteropathogens to environmental enteropathy: findings from the MAL-ED birth cohort study. *EBioMedicine* **18**, 109–117.

Mbuya, M. N. N. & Humphrey, J. H. 2016 Preventing environmental enteric dysfunction through improved water, sanitation and hygiene: an opportunity for stunting reduction in developing countries. *Maternal & Child Nutrition* **12** (S1), 106–120.

Orgill-Meyer, J. & Pattanayak, S. K. 2020 Improved sanitation increases long-term cognitive test scores. *World Development* **132**, 104975.

Ortiz-Correa, J., Filho, M. R. & Dinar, A. 2016 Impact of access to water and sanitation services on educational attainment. *Water Resources and Economics* **14**, 31–43.

Park, M. J., Laksono, B., Sadler, R., Clements, A. & Stewart, D. 2015 Household latrines to control environmental contamination and helminthiasis: an exploratory study in Indonesia. *International Journal of Social Science and Humanity* **5** (5), 429–435.

Petri, W. A., Miller, M., Binder, H., Levine, M., Dillingham, R. & Guerrant, R. 2008 Enteric infections, diarrhea, and their impact on function and development. *The Journal of Clinical Investigation* **118** (4).

Petri, W. A., Naylor, C. & Haque, R. 2014 Environmental enteropathy and malnutrition: do we know enough to intervene? *BMC Medicine* **12** (1), 187.

Pinkerton, R., Oria, R., Lima, A., Rogawski, E., Oria, M., Patrick, P., Moore, S., Wiseman, B., Niehaus, M. & Guerrant, R. 2016 Early childhood diarrhea predicts cognitive delays in later childhood independently of malnutrition. *American Journal of Tropical Medicine and Hygiene* **95** (5), 1004–1010.

Rauniyar, G., Orbeta, A. & Sugiyarto, G. 2011 Impact of water supply and sanitation assistance on human welfare in rural Pakistan. *Journal of Development Effectiveness* **3** (1), 62–102.

Sclar, G. D., Garn, J. V., Penakalapati, G., Alexander, K. T., Krauss, J., Freeman, M. C., Boisson, S., Medlicott, K. O. & Clasen, T. 2017 Effects of sanitation on cognitive development and school absence: a systematic review. *International Journal of Hygiene and Environmental Health* **220**, 917–927. http://dx.doi.org/10.1016/j.ijheh.2017.06.010.

Spears, D. & Lamba, S. 2016 Effects of early-life exposure to sanitation on childhood cognitive skills: evidence from India's total sanitation campaign. *Journal of Human Resources* **51** (2), 298–327.

Trinies, V., Garn, J. V., Chang, H. H. & Freeman, M. C. 2016 The impact of a school-based water, sanitation, and hygiene program on absenteeism, diarrhea, and respiratory infection: a matched-control trial in Mali. *American Journal of Tropical Medicine and Hygiene* **94** (6), 1418–1425.

UNICEF 2018 UNICEF's game plan to end open defecation. Visit: https://www.unicef.org/wash/files/UNICEF_Game_plan_to_end_open_defecation_2018.pdf (Accessed 11 June 2020).

Vyas, S. & Kumaranayake, L. 2006 Constructing socio-economic status indices: how to use principal components analysis. *Health Policy and Planning* **21** (6), 459–468.

Watanabe, K. & Petri Jr, W. A.. 2016 Environmental enteropathy: elusive but significant subclinical abnormalities in developing countries. *EBioMedicine* **10**, 25–32.

Watson, T. 2006 Public health investments and the infant mortality gap: evidence from federal sanitation interventions on U.S. Indian reservations. *Journal of Public Economics* **90** (8–9), 1537–1560.

Wolf, J., Pruss-Ustun, A., Cumming, O., Bartram, J., Bonjour, S., Cairncross, S., Clasen, T., Colford Jr., J., Curtis, V., De France, J., Fewtrell, L., Freeman, M. C., Gordon, B., Hunter, P.R., Jeandron, A., Johnston, R.B., Mausezahl, D., Mathers, C., Neira, M. & Higgins, J. P. T. 2014 Assessing the impact of drinking water and sanitation on diarrhoeal disease in low- and middle-income settings: systematic review and meta-regression. *Tropical Medicine & International Health* **19** (8), 928–942.

First received 4 March 2020; accepted in revised form 19 May 2020. Available online 1 June 2020

Research Paper

Sanitation dynamics: toilet acquisition and its economic and social implications in rural and urban contexts

Britta Augsburg and Paul Rodríguez-Lesmes

ABSTRACT

This paper uses primary micro-data from Indian households residing in rural villages and poor urban neighbourhoods to shed light on household sanitation decision-making. We use a theoretical economic model to reduce the dimensionality and complexity of this process. Beyond the most commonly analysed motivator, health, we consider economic and non-pecuniary benefits. We provide empirical evidence that each of these margins matter, and do so in both rural and urban contexts, and discuss how our findings can be explored in sanitation policy and programme design.

Key words | economic drivers, household investment, India, sanitation, WASH

Britta Augsburg (corresponding author)
Institute for Fiscal Studies,
7 Ridgmount Street, London WC1E 7AE,
UK
E-mail: *britta_a@ifs.org.uk*

Paul Rodríguez-Lesmes†
Universidad del Rosario,
Calle 12 C No. 4, 69 Casa Pedro Fermín de Vargas,
111711 Bogotá, DC,
Colombia

†This research was made possible through the generous support of DGIS as well as ESRC–DfID, award number ES/J009253/1. We want to thank participants at various conferences and workshops, as well as anonymous referees, for their valuable comments, which helped us improve this study significantly. We would also like to thank the FINISH team for their invaluable support throughout the study.

HIGHLIGHTS

- Health is but one motivating factor for households to invest in sanitation in the context of India.
- Households that own sanitation exhibit higher consumption expenditures.
- An increase in productive asset ownership and a shift in time allocation are potential drivers behind increased consumption expenditures.
- Sanitation shifts children's time allocation within the household away from domestic chores and collection of water.
- Sanitation seems to serve as a pre-marital investment strategy.

INTRODUCTION

Safe sanitation, a means of isolating human waste, has been recognised as an indispensable element of disease prevention and primary health care programmes (e.g. the Declaration of Alma-Ata, 1978). The worldwide consensus of its importance led to 'access to adequate and equitable sanitation' becoming part of the Sustainable Development Goals (UN 2015). Yet, with an estimated 1.3 billion people lacking basic sanitation, the scale of the problem is huge (Mara & Evans 2018).

An important challenge to increasing sanitation coverage is its costly provision. According to the World Bank, an estimated US$19.5 billion a year is needed globally to meet nationally defined WASH targets (Hutton & Varughese 2016). An under-acknowledged contributor to investments is households themselves: based on survey responses by 35 national governments in 2018/2019, households contribute an estimated 66% of US$52 billion of annual WASH expenditures (WHO 2019) These figures have triggered calls for a stronger emphasis on research that enables a better understanding of household investment in WASH (Danert & Hutton 2020). Novotný *et al.* (2018) highlight that research aimed at understanding how contextual factors and

doi: 10.2166/washdev.2020.098

motivations affect different sanitation outcomes is currently underdeveloped and that the current programmatic focus provides a narrow understanding of sanitation dynamics.

In this paper, we respond to this identified gap. We use primary micro-data from households residing in two Indian states to shed light on household sanitation decision-making, exploring the association of household characteristics with revealed preference for toilet uptake as well as outcomes resulting from the acquisition choice. Since such sanitation dynamics are characterised by complex human–environment interactions (Novotný *et al.* 2018), we present and structure our analysis around a theoretical economic model, which helps to reduce dimensionality and complexity. In addition to health and non-pecuniary benefits, the model highlights the importance of economic factors as motivators, a category that Novotný *et al.* (2018) identify as under-represented in the sanitation literature.

Our data include two survey rounds implemented in both rural villages and poor urban neighbourhoods, allowing us to provide a rich picture of the main correlates with, and potential outcomes of, sanitation uptake in different environmental contexts. We clarify the distinct socio-economic-cultural contexts and resulting differences in motivations for, and impacts of, toilet construction throughout our analysis but do not intend to use aggregates to inform policy or programming.

India is a particularly apt context to study household investment in sanitation, having contributed over 50% of the close to 700 million people who defecate in the open globally in 2017 (UNICEF and WHO 2019). The Government of India has shown a significant commitment to achieving SDG 6 of clean water and sanitation for all by 2030, including its ambitious Clean India Movement.

METHODS

Data and study population

We use data collected as part of an evaluation effort for a sanitation intervention. The original study's baseline and endline reports (Augsburg & Rodríguez-Lesmes 2015) provide detailed information on the selection of study communities and respondents and on survey instruments, attrition, etc. Data collection followed *The Netherlands Code of Conduct for Scientific Practice*. Clearance to collect data was given by the United Nations University-MERIT, The Netherlands. Approval for the second round in Tamil Nadu was separately given by UCL IRB (project code 2168/010). Two rounds of data were collected in 2009/2010 and 2013/2014. We include in our analysis those households that were interviewed twice, allowing us to apply panel data models with household fixed effects in our analysis. We worked with a sample of 1,035 households in 39 poor urban neighbourhoods, 764 households in 17 peripheral rural villages of Gwalior city, Madhya Pradesh and 869 households in 46 GPs (GPs constitute the smallest administrative unit in India) in Thiruvarur, Tamil Nadu (henceforth we refer to peripheral villages and GPs combined as 'rural villages/areas' and to the poor urban neighbourhoods, which are characterised by substandard housing and infrastructure, as 'slums'). Appendix A1 provides further information on survey locations and data collection (timings and sample sizes). The main survey instruments were a general household survey and an individual survey with the head of household, if female, or the spouse, if male and married (whom we refer to as the 'main woman'). Apart from household general characteristics, the survey instruments include detailed information on living standards, assets, consumption expenditures (including 21 food items), income, risk perceptions, credits, savings and insurance and demand for health care. A distinguishing feature of the data is an extensive module on sanitation and hygiene facilities, practices and perceptions. The interview with the main woman of the household covered information on time utilisation, hygiene practices and knowledge, cultural background and measures of empowerment. The women were also asked about children in their household, particularly providing information on their health status, time utilisation and nutrition. Survey questions were consistent across contexts with respect to variables used in this analysis, with response options covering both contexts; the Gwalior survey included additional modules on time use and detailed distance data.

Table 1 provides a set of descriptive statistics of study households. Descriptive statistics on further variables, including outcomes considered in the subsequent analysis,

Table 1 | Descriptive statistics of study households

		Round 1		Round 2	
		Rural	'Slums'	Rural	'Slums'
Religion & caste					
Muslim		9.9%	25.8%	9.8%	25.8%
Hindu		87.7%	74.0%	87.7%	74.0%
Forward caste		12.0%	30.8%	12.3%	30.8%
Most backward caste		12.7%	7.3%	12.7%	7.3%
Backward caste		46.1%	35.0%	45.3%	35.0%
Scheduled caste		27.1%	24.8%	27.7%	24.8%
Scheduled tribe		2.1%	2.2%	2.0%	2.2%
Household (HH) demographic composition					
No. of HH members		5.1	5.4	5.3	5.9
No. of male HH members		2.7	2.9	2.7	3.0
HH has at least one child <6 years old		30.9%	37.1%	30.9%	39.1%
Main woman: married		91.6%	91.7%	91.5%	88.8%
Main woman: age		37.8	36.6	38.1	37.98
Main woman: no formal education		44.7%	56.1%	45.6%	55.0%
Dwelling characteristics & location					
Dwelling is owned		93.6%	83.4%	94.9%	86.6%
Dwelling has a bathroom		38.2%	70.0%	63.8%	78.9%
Pucca (strong)		23.9%	40.3%	50.4%	75.1%
Semi-pucca (semi-strong)		41.0%	46.9%	23.4%	18.6%
Value of the dwelling (INR 1,000)		142.1	184.1	265.4	245.6
In Gwalior area		46.8%	100.00%	46.8%	100.00%
'Slum'		0.00%	100.00%	0.00%	100.00%
Sanitation & hygiene					
Main source of drinking water is HH service connection		16.5%	29.8%	24.1%	36.5%
HH has a toilet		24.6%	53.7%	42.0%	71.7%
Type of toilet:	Pour/flush to pit or septic tank	88.5%	40.8%	77.7%	42.2%
	Pour/flush to drainage	0.7%	14.0%	0.3%	13.2%
	Pour/flush to other	2.0%	31.2%	5.2%	24.9%
	Simple pit	7.0%	3.2%	16.2%	19.3%
	Don't know/No answer	1.7%	10.8%	0.6%	0.4%

Notes: Means for rounds 1 and 2, in rural and 'slum' areas. 'Rural' is peripheral villages of Gwalior and villages in rural Tamil Nadu; 'slums' are in the city of Gwalior. 'Value of the dwelling' is in 1,000 Indian Rupees of 2013 (i.e. round 1 numbers were adjusted by a factor of 1.32, which was calculated based on national figures for 2011, 2012 and 2013).

are provided in Appendix A2. Most households are Hindu (83%) and about 16% are Muslim; 19% report belonging to forward castes, 42% to backward castes, 26% to scheduled castes, 2% to scheduled tribes and 11% to most backward castes. A typical household comprises five members, three of whom are male, and around 33% have at least one child under the age of 6 years. The main woman in the household was on average 37 years of age and married (91%), and almost half (49%) had no formal education. Compared with rural villages, urban 'slums' are characterised by a higher percentage of Muslim and forward caste households. Households are slightly larger and the main woman less educated.

Dwelling ownership rates are lower, but dwellings are of stronger structure and of higher value.

Households had an average income of US$0.97 per member per day, considerably below the commonly used international poverty line of US$1.25. About a fifth of households had piped water access. A bit more than one-third of the sample has a toilet and almost half a bathroom. Toilet ownership was verified by the interviewer, alleviating concerns that households misreport their sanitation ownership status, possibly due to embarrassment about not having sanitation facilities. Verification was not possible for 6% of the sample, in remaining cases, respondents' reports were confirmed in 95–98% of cases. Toilet ownership rates varied considerably across study locations, being significantly higher in urban 'slums' (54%) than rural (25%) areas. Ownership rates increased significantly over time, with a bit more than half of study households owning a toilet after about three years (72% for 'slums', 42% for rural). In both survey rounds, pour/flush to pit or septic tanks were the most common toilet types available in rural settings. In contrast, pour/flush to drainage/others were slightly more common types in urban settings. Simple pit toilets became more common in round 2, especially in 'slums'.

Theoretical framework

In order to understand how households make sanitation investment decisions within this context, and the consequences these decisions might have, we present a simple theoretical model. The model is developed in detail in Appendix A3. Here we present its main elements and implications that link to our empirical analysis.

Within the simple two-period framework we consider, households care about consumption, leisure and health. They must decide who works and for how long, how many resources to consume today, how much to save for the future and whether to invest in a toilet today. Toilets are modelled as a one-off durable and indivisible investment, which can be expensive relative to household income. In the context of this study, households spend up to 20% of their annual income to construct toilets. Households benefit from owning a toilet through three channels. First, toilets add value to the household's dwelling, which can be sold or serve as collateral, hence increasing a household's asset wealth and ability to borrow. Second, toilets increase a household's utility through intangible benefits, such as comfort, status and even marriage market prospects for grooms-to-be. Third, toilets improve future health, which the household values in itself and which furthermore increases household productivity. While there is, to date, no clear evidence on the link between sanitation ownership and productivity, health is increasingly seen as an important part of human capital. An active literature demonstrates the link between health and human capital and productivity (Strauss & Thomas 1998; Currie & Hyson 1999; Currie & Madrian 1999; Cunha & Heckman 2007). The model also incorporates health externalities, i.e. that health benefits derived from improved sanitation depend not only on a household's own ownership status but also on the status of its neighbours (Kresch *et al.* 2020).

This set-up guides our empirical analysis, which is split into two parts. For the first part – an analysis of toilet acquisition – we consider, for example, the role of the household's current wealth/income, sanitation ownership of its neighbours and proxies for anticipated marriage decisions. For the second part of our empirical analysis – understanding potential benefits of owning a toilet – the model provides testable hypotheses for households that are not liquidity constrained and hence are able to make the sanitation investment. For these households, the investment moves resources from the present into the future. As a result, according to the model, we should observe an increase in the value of assets owned, either because the investment itself adds value or because additional assets are purchased. This part of the model leads us to look at the information on owned assets as outcomes. As mentioned above, such an increase in assets might allow households to obtain access to credit markets, which we consider through information on household borrowing. Additionally, improved health and productivity (potentially in combination with productive investments) can increase household income. Changes in the household's permanent income and wealth have, in turn, implications for consumption patterns, leading us to look at households' consumption expenditures. Finally, sanitation investment might affect the balance between consumption and labour in two ways: via productivity or via the marginal utility of labour and consumption, induced directly through the impact on health. This implies that the impact of sanitation investment on labour supply is ambiguous. We will bring it to the data through information on children's time use. The analysis section will discuss the full set of outcomes that we consider.

Finally, although households in this model can borrow and save, some might remain liquidity constrained given imperfect financial markets with borrowing limits that depend on a household's wealth, and hence only indirectly on its income potential (Gautam 2018).

Empirical strategy

Our analysis is structured around two main objectives of (1) assessing determinants (correlates) of toilet acquisition and (2) understanding the potential benefits of toilet ownership for several outcomes.

We analyse *correlates of toilet acquisition* through a logistic regression where we constrain our sample to households that had no toilet at the time of the first survey round. It establishes the conditional correlation between a set of covariates X and toilet ownership status T.

$$T_{i,j,t=2} = g(X'_{i,j,t=1}\,\beta_1 + u_{i,j,t=2}) \tag{1}$$

Variables are at the level of the household i, the community j and time t, with $t=\{1,2\}$ representing the two survey rounds. The vector of estimated parameters $\hat{\beta}_1$ gives us an idea of the correlation between each variable on the right-hand side and toilet ownership. We cluster the error term u at the community level. We run this regression for the entire sample, as well as separately for rural and urban 'slum' areas. We confirm the robustness of our findings by, first, assessing the sensitivity of estimates to the inclusion of a different set of covariates, X, and, second, using a linear instead of a logistic regression model.

For our analysis of the *link between toilet ownership and outcomes*, we can consider potential impacts by regressing outcomes on toilet ownership, conditional on the above determinants, which include confounding factors, such as income and education. We use a linear panel model, which includes a time (survey round) dummy (γ_t) and household fixed effects (α_i).

$$Y_{ijt} = \delta\, T_{ijt} + X'_{ijt}\omega_2 + \alpha_i + \gamma_t + u_{ijt} \tag{2}$$

Other variables and subscripts are defined as above. Monetary outcomes are transformed using the inverse hyperbolic sine transformation (IHS), which is similar to a logarithmic transformation but allows for zeros in its domain. The resulting coefficients $\hat{\delta}$ and $\hat{\omega}_2$ are to be interpreted as semi-elasticities. While this model eliminates time-invariant confounding factors, we avoid overstating the substantive importance of the variable's effect. In addition to potential omitted time-*varying* variable bias, another point to be aware of is that two rounds of data do not allow us to analyse sequencing. For example, investing in a toilet could improve the asset value of the house, allowing households to get more resources in the financial market. As a result, households can invest in productive assets and increase their income and consumption. However, toilets could be the result of an unobserved shock that increased the household's income, which allows it to invest both in toilets and in other assets. To get a step closer to establishing causality, we estimate a difference-in-difference linear probability model on a subsample of households matched based on the probability that they will construct a toilet in the future (Rosenbaum & Rubin 1983; Heckman *et al.* 1997) for further discussion on propensity score matching (PSM). The procedure was implemented using kernel matching on the propensity score with the statistical software Stata (Leuven & Sianesi 2014).). Although findings are in line (as discussed later), Equation (2) remains our preferred specification since the matching exercise results in a significantly reduced sample size.

RESULTS

Correlates of toilet acquisition

Table 2 presents the results of the analysis of determinants of toilet acquisition. Our sample is households that changed toilet ownership status from having no toilet at survey round 1 to having one at round 2, 3–4 years later. These

Table 2 | Toilet acquisition determinants

Independent variable *X*	Base mean (1) Rural	(2) 'Slums'	(3) Toilet	Average marginal effects (4) All	(5) Rural	(6) 'Slums'	(7) Difference
Household income (past year) per household member							
Belongs to quartile 2 or above	74.2%	69.2%	39.8%	3.32	−0.01	10.54**	8.81*
				(2.51)	(2.67)	(5.34)	(5.35)
Belongs to quartile 3 or above	44.9%	37.5%	46.7%	7.10**	8.95***	4.63	−5.86
				(2.92)	(3.42)	(6.07)	(6.23)
Belongs to quartile 4	19.4%	13.5%	55.5%	−1.91	−0.86	−9.71	−7.17
				(3.75)	(4.21)	(8.44)	(8.40)
Social background							
Religion: Muslim	10.0%	28.4%	44.9%	−6.41	−5.55	−3.50	3.13
				(4.83)	(5.59)	(8.47)	(9.30)
Most backward caste	12.1%	30.6%	29.5%	6.60	11.11	4.23	−8.60
				(4.93)	(7.22)	(7.87)	(10.16)
Backward caste	41.3%	8.2%	43.4%	0.64	7.02	−7.91	−14.27
				(5.52)	(7.73)	(8.83)	(11.17)
Scheduled caste or tribe	34.8%	29.0%	22.0%	−14.91***	−8.75	−19.86**	−7.04
				(5.51)	(7.78)	(8.44)	(11.23)
Household demographic composition							
Number of members	5.0	5.4		1.22	2.76**	−2.23	−4.87*
				(1.23)	(1.25)	(2.91)	(2.77)
Number of male members	2.7	2.9		−4.28**	−4.93***	−2.12	3.61
				(1.66)	(1.79)	(3.46)	(3.44)
HH has at least one child under 6 years old	32.0%	40.4%	38.1%	−1.58	−2.74	3.33	5.77
				(2.84)	(3.56)	(5.02)	(5.68)
There is an unmarried boy aged 17–24 years	33.4%	28.6%	39.7%	9.57***	7.75***	12.06	1.61
				(2.60)	(2.25)	(7.88)	(7.04)
There is an unmarried girl aged 13–20 years	32.6%	31.3%	38.2%	−4.63	−5.41	−0.22	5.72
				(3.56)	(3.73)	(7.97)	(7.72)
Main woman completed grade VI, VII or VIII	15.0%	9.5%	39.6%	3.93	5.36	3.20	−3.18
				(3.74)	(3.90)	(9.16)	(8.65)
Main woman completed grade IX or above	11.3%	5.1%	48.1%	3.94	4.32	4.14	−1.26
				(4.77)	(5.16)	(13.38)	(12.44)
Sanitation reported as a decision factor for marriage	26.3%	39.2%	50.9%	4.28	−1.07	11.76**	11.00**
				(2.71)	(2.79)	(4.76)	(5.09)
Dwelling characteristics							
Strong structure (pucca)	18.0%	22.8%	59.8%	5.51	4.88	4.80	−1.33
				(3.50)	(4.03)	(7.08)	(7.32)
Semi-strong structure (semi-pucca)	41.6%	57.0%	35.5%	7.24**	6.69**	7.67	−0.89
				(2.84)	(3.20)	(5.42)	(5.77)

(continued)

Table 2 | continued

Independent variable *X*	Base mean (1) Rural	(2) 'Slums'	(3) Toilet	Average marginal effects (4) All	(5) Rural	(6) 'Slums'	(7) Difference
Community characteristics							
Coverage: water service connections	5.1%	19.5%		−0.22*	−0.43**	−0.20	0.30
				(0.13)	(0.20)	(0.14)	(0.25)
Coverage: toilet ownership	8.8%	38.2%		0.52***	0.79***	0.49***	−0.45
				(0.12)	(0.22)	(0.14)	(0.28)
Total number of HHs surveyed in the village/'slum'	52.7	53.3		0.14**	−0.15***	0.06	−0.12
				(0.05)	(0.06)	(0.05)	(0.08)
In Gwalior area	51.7%	100.0%	54.0%	−15.86**	−16.47*		
				(6.42)	(8.47)		
Total observations				1,509	1,058	451	1,509
Total villages/'slums' (clusters)				96	62	34	96

Notes: Columns 1 and 2 present the mean at survey round 1, in rural and 'slum' areas, respectively, of each of the independent variables in the rows for households that did not have a toilet in their home at the first round (base mean). 'Rural' is peripheral villages of Gwalior and villages in rural Tamil Nadu; 'slums' are in the city of Gwalior. As a reference, column 3 presents the percentage of households that own a toilet conditional on the specific category (for discrete variables only). Columns 4–7 correspond to average marginal effects (AME), for each of the independent variables, after logistic regressions where the dependent variable is having a toilet in the house at the second round, conditional on not having one in the first round. Column 4 is for the entire sample, and columns 5 and 6 for rural areas and 'slums', respectively. Column 7 presents the difference between the estimates in columns 5 and 6, obtained by interacting 'toilet ownership' (as well as all other independent variables) with the 'slum' indicator. The four regressions include as controls all the variables presented in the table plus: a binary variable for the presence of any major shock to the household in the last 12 months; the age of the woman who is the household head or the spouse of the household head; an indicator of whether she lives with her in-laws; an indicator of whether the dwelling is owned by the household; and the value of household elements. Standard errors, clustered at the village/'slum' level, are shown in parentheses. Significance: *10%, **5%, ***1%.

are 17% of the rural and 18% of our 'slum' sample. The majority of these newly acquired toilets were funded through households' own money/savings (80%), followed by government subsidies (8%) and informal loans (9%); remaining categories are formal loans (1%) and 'other' (2%). The data reveal a clear wealth gradient in the type of funding used by households, with richer households more likely to use their own money/savings to construct a toilet (highest income quartile (Q4) 90%; 'slums': 92%; rural: 89%) and poorer households more likely to get government support (Q1 13%; 'slums': 21%; rural: 10%) and informal loans (Q1 12%; 'slums': 11%; rural 14%). Columns 1 and 2 provide sample averages of the considered variable for the subsample of households without a toilet in the first survey round in rural and urban 'slum' areas, respectively. As a reference for the analysis, column 3 gives the average toilet ownership rate at round 1 for the respective variable, if binary. Columns 4, 5 and 6 present our estimated coefficients of interest, $\hat{\beta}_1$, as specified in Equation (1). Column 7 presents the difference between the estimates in columns 5 and 6, obtained by interacting toilet ownership (as well as all other independent variables) with the 'slum' indicator.

The first two panels provide results on wealth and status determinants. We observe a positive gradient of toilet ownership and social standing: the means of toilet ownership presented in column 3 show that households with higher income and higher caste are more likely to own a toilet. In terms of acquisition, our results reveal that those from the third income quartile are significantly more likely to invest in a toilet between the two survey rounds. Rural areas, where toilet ownership rates are generally much lower than in urban 'slums', drive this result. Interestingly, we also find that in 'slums', it is those in the second income quartile or above that are significantly more likely to acquire a toilet over time. At the same time, households in a scheduled caste or tribe are less likely to improve their sanitation ownership than those in a forward caste – a finding that is more pronounced in 'slums' than in rural areas.

Correlations of toilet acquisition with other proxies of household wealth are in line with those of income:

households with dwellings of semi-strong (rather than strong or weak) structures are significantly more likely to invest in a toilet between the two survey rounds, again driven by rural areas.

We further find that certain household compositions and changes are driving sanitation investments. Specifically, the arrival of a new female (an increase in household size, conditional on the number of males) in a rural household increases the likelihood of constructing a toilet significantly, a finding not driven by a newborn member. Further, a male close to the legal marriage age (21) in the household makes investment more likely.

Finally, we find that conditional on contextual characteristics, such as wealth or income, an increase in sanitation coverage is positively associated with greater sanitation uptake in the village, as predicted by the theoretical model. We also find a negative association between the coverage of water service connections in rural villages and toilet ownership, in line with Bennett's (2011) results, which are consistent with the hypothesis that clean water can serve as a substitute for sanitation.

Toilet ownership and outcomes

Table 3 shows estimates of parameter δ (Equation (2)) in columns 4–6. Column 7 shows the difference between rural and 'slum' associations, similar to column 7 of Table 2. Inspired by the model, we group outcomes around health, household consumption, wealth and finances and time allocation.

Health

Health is typically considered a key motivator for improved sanitation and also highlighted in our model as one margin along which households might benefit. Within this analysis, we do not find evidence for reduced illness associated with sanitation, measured by diarrhoea incidence among children in the last week. We do, however, find evidence that households perceive health benefits of owning a toilet, particularly in urban 'slums', where the main respondent is 11 percentage points more likely to perceive herself as healthier than peers in the community and 10 percentage points more likely to perceive her family as healthier than other families in the community (significant at the 5 and 1% level, respectively).

Consumption

The model predicts that toilet ownership can result in increased income. Since income is an important confounding factor, which we account for throughout our analysis, we consider consumption expenditure, rather than income, as an outcome variable. A change in consumption expenditures is typically a result of a permanent change in income (Jappelli & Pistaferri 2010) and hence a valid proxy. We find a large, positive and significant correlation between sanitation ownership and household consumption, observed in rural and urban 'slum' areas alike. Particularly, non-durable consumption (such as transport, utilities, fuel, education and cosmetics) is higher for those with a toilet. We stress that this relationship holds despite accounting for income in our analysis, which in fact could downward-bias estimates. The higher expenditures are suggestive that households with a toilet are more productive, potentially driven by better health, ownership of more productive assets or differing time allocation.

Wealth and finances

And indeed, we find significant relationships between sanitation and likely productive assets the households own, including vehicles (bicycle, scooter, motorbike, four-wheeler) and other household items. We also observe an increase in the value of agricultural assets for 'slum' households. Further analysis reveals that these assets are uncommon and small, with the results driven by about 2% of the sample.

We further find that sanitation investments are reflected in the value of the dwelling, much above the investment needed to construct the toilet. The average reported value of the toilet is INR 20,000, whereas the estimated coefficient suggests an increase in value of INR 42,926; houses without a toilet were on average worth INR 103,900 in 2013. We acknowledge that housing valuations are notoriously difficult in the setting and self-reported, but even if misreported, a perceived increase in value would be an important investment motivation.

Table 3 | Toilet ownership and outcomes

	Base mean		Full sample linear fixed effects panel		Linear fixed effects panel rural versus 'slums'		
Outcome variables *Y*	**(1) Rural**	**(2) 'Slums'**	**(3) Obs.**	**(4) Coeff.**	**(5) Rural**	**(6) 'Slums'**	**(7) Difference**
Health							
At least one child had diarrhoea last week	7.3%	13.2%	1,732	1.44	1.42	1.14	−2.33
				(2.50)	(5.14)	(2.90)	(6.30)
Perceives herself as healthier than others	33.8%	23.8%	1,709	9.18**	5.63	11.27**	7.78
				(3.72)	(6.77)	(4.26)	(9.67)
Perceives family as healthier than others	30.8%	21.6%	1,701	7.13	1.33	10.36***	11.53
				(4.50)	(10.13)	(3.75)	(11.74)
Annual consumption (previous year) per household member (1,000 INR)							
(IHS) Total consumption expenditures	91.4	96.3	2,505	14.60%***	15.90%**	13.09%***	−6.20%
				(3.94)	(6.81)	(3.62)	(8.10)
(IHS) Food consumption	56.6	74.4	2,505	5.48%	6.70%	4.89%	18.52%
				(6.56)	(11.54)	(3.85)	(11.61)
(IHS) Non-durable consumption	23.2	16.8	2,496	22.05%***	25.29%***	18.60%**	−14.69%
				(5.07)	(7.09)	(7.60)	(10.21)
(IHS) Alcohol and tobacco consumption	0.8	0.9	2,505	−4.05%	−8.93%	1.41%***	−5.61%
				(3.67)	(6.51)	(0.46)	(6.25)
Household asset values (INR 1,000)							
(IHS) Dwelling	104.3	103.9	1,980	41.34%***	46.67%***	34.98%**	−29.03%
				(8.17)	(10.30)	(13.67)	(19.17)
(IHS) Vehicles	5.2	3.0	2,505	38.92%***	43.47%***	32.63%***	−12.36%
				(8.00)	(12.26)	(9.81)	(15.27)
(IHS) Household items	47.0	55.9	2,505	39.81%***	56.02%***	16.91%*	−57.38%***
				(9.35)	(15.87)	(9.15)	(19.49)
(IHS) Agricultural assets	25.3	6.8	2,505	10.16%	6.95%	15.47%**	−17.24%
				(6.66)	(10.52)	(7.52)	(12.95)
Credit							
Household has debt outstanding	43.1%	22.2%	1,972	−0.45	−3.19	3.39	10.06
				(4.11)	(4.57)	(7.16)	(8.13)
(IHS) Amount of debt outstanding (INR 1,000)	14.5	5.4	1,884	20.18%	30.45%	2.60%	−66.83%*
				(17.77)	(27.60)	(25.54)	(40.22)
Children aged 3–15, time allocation (Gwalior only) – % with more than 0 h							
Domestic housework	47.3%	52.1%	1,266	−11.84***	−5.50	−14.64***	−15.49*
				(3.98)	(7.35)	(3.38)	(8.14)
Carrying water	46.2%	49.6%	1,259	−15.18**	−12.82*	−15.19***	−12.60
				(5.72)	(6.00)	(7.07)	(8.46)
Work in household business	4.9%	1.8%	1,258	−1.28	−2.62	0.78	−6.43
				(2.29)	(4.51)	(2.08)	(5.52)

(*continued*)

Table 3 | continued

Outcome variables *Y*	Base mean		Full sample linear fixed effects panel		Linear fixed effects panel rural versus 'slums'		
	(1) Rural	(2) 'Slums'	(3) Obs.	(4) Coeff.	(5) Rural	(6) 'Slums'	(7) Difference
Play	83.2%	83.2%	1,260	2.04	−3.79	6.45*	2.69
				(2.67)	(3.65)	(3.23)	(5.06)
Taking care of elders or sick household members	43.0%	45.4%	1,261	−1.82	−2.32	−0.73	−7.59
				(4.95)	(6.72)	(6.45)	(8.56)
Extracurricular study	13.6%	14.8%	1,266	2.81	−1.22	5.14	−2.71
				(3.84)	(6.86)	(4.83)	(8.81)

Notes: Each row represents the dependent variable of a set of fixed effects linear panel models. Columns 1 and 2 present the mean at survey round 1, in rural and 'slum' areas, respectively, of each of the dependent variables in the rows, for households that did not have a toilet in their home at the first round (base mean). 'Rural' is peripheral villages of Gwalior and villages in rural Tamil Nadu; 'slums' are in the city of Gwalior. Column 3 presents the number of observations of the regression in the 'full' sample. Columns 4, 5 and 6 present the coefficient estimate for the independent variable 'toilet ownership' when considering the full, rural and 'slums' sample, respectively. Column 7 presents the difference between the estimates in columns 5 and 6, obtained by interacting 'toilet ownership' (as well as the other independent variables) with the 'slum' indicator. All regressions include as controls: log of income, the presence of any major shock to the household in the last 12 months, the numbers of total and male household members, dummies for the presence of marriageable-age-male/marriageable-age-female/under-5 members, and the proportion of water that comes from piped drinking water in the community. Consumption expenditure, amount of debt and asset values are in 1,000 Indian Rupees of 2013 (i.e. round 1 numbers were adjusted by a factor of 1.32, which was calculated based on national figures for 2011, 2012 and 2013). For the regressions, these values were transformed using the inverse hyperbolic sine (IHS), which for interpretation is similar to the logarithmic transformation. Standard errors, clustered at the village/'slum' level, are shown in parentheses. Significance: *10%, **5%, ***1%.

Interestingly, the significant associations of toilet ownership and consumption and wealth are relevant in both rural and 'slum' contexts but are generally quantitatively larger in rural communities.

Finally, results on credit outcomes suggest no changes in borrowing on either the extensive or intensive margin.

Time allocation

Our model predicts a potential change in labour supply. We do not have the relevant data to analyse this prediction. We can, however, check whether time allocation within the household differs depending on toilet ownership. In particular, we have data on the time use of children aged 3–15 years in Gwalior (adult time-use data are not available and no time use was collected in Tamil Nadu). The last panel of Table 3 shows associations between toilet ownership and indicators for a positive amount of time reported on a specific activity by at least one child within the considered age range when data were collected 3–4 years after the base year. As a robustness check, we also estimated the relationship with the reported average hours of all the children in the household. The findings are in line. We find evidence that especially children living in urban 'slums' are significantly less likely to spend time on domestic housework and carrying water and instead more time on play when they live in a dwelling with a toilet. Children living in households with a toilet in rural areas are less likely to spend time carrying water.

Robustness checks: matching

As stressed previously, while the presented data account for a large set of covariates and we consider consistency across time and context, results might still be driven by pre-existing trends. In other words, households that were already improving along several margins for reasons that we cannot observe might acquire toilets. We, therefore, employ as a robustness check a PSM exercise, which ensures we compare households that were as similar as possible along observable characteristics. Appendix A4 presents evidence of a successful PSM procedure (Figure A2), providing us with a comparable sample. We find that our main conclusions hold true and also that cross-sectional analysis with both survey rounds confirms our conclusions: coefficients remain in line (Table A3), but at times with reduced significance driven by the reduced sample (the matching procedure reduced the analysis sample by almost half) and hence larger standard errors. We lose significance altogether for the perceived health of the family in urban 'slums', consumption and the value of transport assets in rural settings, dwelling value, the value of household items and some time-use variables.

DISCUSSION

The theoretical model has provided a useful foundation to support our analysis. Here, we use it to guide the discussion of our results and explain their broader relevance. We start with the motivating factor of health for sanitation investment.

While our analysis does not provide evidence of improved health associated with *private* toilet ownership, we know from Augsburg & Rodríguez-Lesmes (2018) that children's height for age improves in our Gwalior sample with higher *community* sanitation coverage. The strong association with health perceptions present in our data suggests that households internalise such health impacts, which have also been shown in other studies (see, for example, Prüss-Ustün *et al.* (2014), Pickering *et al.* (2015) or, for an overview, UN (2016)). The finding that these are primarily driven by 'slum' dwellers' perceptions is possibly a result of higher population density in these areas, and hence higher vulnerability to negative externalities of poor sanitation, as discussed in Hathi *et al.* (2017). Our findings show that toilet acquisition increases with community toilet coverage. Potential drivers of this finding include that households internalise the sanitation externality (Guiteras *et al.* 2019), or social network effects (Shakya *et al.* 2015).

Our results provide further contributions to this literature stream conducting benefit analysis of improved sanitation (see, for example, Hutton & Haller 2004) in three ways. For one, we are able to document an increase in consumption expenditures, which provides empirical evidence for the hypothesis that sanitation ownership can induce an income effect. Second, we find that households that own a toilet also acquire more other assets than households without a toilet. Some of these are likely productive (such as transport), a potential driver behind higher consumption expenditures. Others can be once again linked to social status and/or non-pecuniary benefits driving sanitation investment: despite having a similar income, households with sanitation might also have a better quality of life in general. Third, recent literature argues that sanitation has an important role to play in the formation of human development, such as cognitive skills and learning (Spears & Lamba 2016; Britto *et al.* 2017; Sclar *et al.* 2017). While in this literature improved health is typically seen as the driving factor, our analysis shows that improved sanitation can lead to children shifting time from household chores to play, and potentially education. Such changes in time use are typically encompassed in calculations of the economic benefits of sanitation. OECD (2011), for example, uses the estimate that improved sanitation gives a household up to 1,000 additional hours a year to work, study, care for children and so on.

Our analysis also reveals further insights into non-pecuniary benefits of sanitation. A body of literature highlights that latrine adoption is motivated by factors, such as convenience, comfort, cleanliness, privacy, safety and prestige (WSP 2004; Pearson & McPhedran 2008). Our finding of higher sanitation investment with the presence of a male close to marriageable age and with the arrival of an adult female member suggests that toilet construction might further serve as a pre-marital investment strategy and/or that the bride-to-be brings additional resources that allow the investment in a toilet to be made.

Finally, while our model suggests that increased wealth can improve access to borrowing, we do not find any statistically significant evidence for this. While we show a catch-up in toilet acquisition along income and caste gradients, our study also reveals that a large percentage of the population, in particular, those at the lower end of the income distribution in rural areas and those of lower caste, do not invest in sanitation. Households' own reporting that toilets are too expensive hints at liquidity constraints being important drivers.

CONCLUSIONS

Gaining a deeper understanding of contextual factors and motivations that affect sanitation outcomes is important given the current push and aim to provide safe sanitation for all, as manifested in Sustainable Development Goal 6.2.

While not experimental, our findings can be used to contextualise interventions better and to find different angles to promote sanitation. The most commonly adopted strategy to combat open defecation, Community-Led Total Sanitation (CLTS), relies on information and collective action mechanisms. This intervention can have substantial impacts on households' sanitation investments (Pickering *et al.* 2015; Briceno *et al.* 2017; Abramovsky *et al.* 2020). Yet, it is not a silver bullet to achieve open-defecation-free status and policymakers working towards meeting the sanitation-related sustainable development goals are advised to consider alternative or complementary policies (Abramovsky *et al.* 2020). The finding that a significant part of the population we consider seems to be lacking funds to make the investment suggests that complementary policies can be a fruitful angle. Programmes, such as microcredit for sanitation (Augsburg *et al.* 2019, 2020) and subsidy provision (Patil *et al.* 2014; Guiteras *et al.* 2015; Andrés *et al.* 2020), have been successful and are worth considering. Subsidy provision is costly, but the productivity and income gains suggested by our results support further resource allocation to sanitation.

It is interesting to note, though, that a significant percentage of the richest households in our context still do not own a toilet at the time of the second survey round (46% in rural areas and 16% in urban 'slum' areas). It is likely that these households do not perceive a high enough benefit from making the investment, rather than them being liquidity constrained. At the same time, Abramovsky *et al.* (2020) show that CLTS is not effective in richer areas, making this a potentially challenging part of the population to motivate for sanitation investment.

Our findings suggest that a policy addressing non-pecuniary benefits, such as the 'No Toilet No Bride' campaign successfully implemented by the Government of Haryana, India (Stopnitzky 2017), might work well in our study context, particularly the urban 'slums'.

Identifying and assessing which policies are effective in further closing the sanitation gap, particularly when working at scale, should be a priority for researchers, policymakers and implementers in the field.

DATA AVAILABILITY STATEMENT

All relevant data are included in the paper or its Supplementary Information.

REFERENCES

Abramovsky, L., Augsburg, B., Lührman, M., Oteiza, F. & Rud, J. P. 2020 *Community Matters: Heterogeneous Impacts of a Sanitation Intervention*. Institute for Fiscal Studies (IFS) Working Paper W19/11. Available from: https://www.ifs.org.uk/publications/14143 (accessed 17 August 2020).

Andrés, L. A., Deb, S., Joseph, G., Larenas, M. I. & Grabinsky Zabludovsky, J. 2020 *A Multiple-Arm, Cluster-Randomized Impact Evaluation of the Clean India (Swachh Bharat) Mission Program in Rural Punjab, India*. World Bank Policy Research Working Paper 9249. Available from: https://openknowledge.worldbank.org/handle/10986/33796 (accessed 17 August 2020).

Augsburg, B. & Rodríguez-Lesmes, P. A. 2015 *FINISH Impact Evaluation Report*. Available from: https://www.ifs.org.uk/publications/8857 (accessed 17 August 2020).

Augsburg, B. & Rodríguez-Lesmes, P. A. 2018 Sanitation and child health in India. *World Development* **107**, 22–39.

Augsburg, B., Caeyers, B. & Malde, B. 2019 *Can Micro-Credit Support Public Health Subsidy Programs?* World Bank Policy Research Working Paper 8846. Available from: https://openknowledge.worldbank.org/handle/10986/31671 (accessed 17 August 2020).

Augsburg, B., Caeyers, B., Giunti, S., Malde, B. & Smets, S. 2020 *Labelled Loans and Human Capital Investments*. Institute for Fiscal Studies (IFS) Working Paper W20/20. Available from: https://www.ifs.org.uk/publications/14921 (accessed 17 August 2020).

Bennett, D. 2011 Does clean water make you dirty? Water supply and sanitation in the Philippines. *Journal of Human Resources* **47** (1), 146–173.

Briceno, B., Coville, A., Gertler, P. & Martinez, S. 2017 Are there synergies from combining hygiene and sanitation promotion campaigns: evidence from a large-scale cluster-randomized trial in rural Tanzania. *PLoS One* **12** (11), e0186228.

Britto, P. R., Lye, S. J., Proulx, K., Yousafzai, A. K., Matthews, S. G., Vaivada, T., Perez-Escamilla, R., Rao, N., Ip, P., Fernald, L. C. H., MacMillan, H., Hanson, M., Wachs, T. D., Yao, H., Yoshikawa, H., Cerezo, A., Leckman, J. F., Bhutta, Z. A. & the Early Childhood Development Interventions Review Group 2017 Nurturing care: promoting early childhood development. *Lancet* **389** (10064), 91–102.

Cunha, F. & Heckman, J. 2007 The technology of skill formation. *American Economic Review* **97** (2), 31–47.

Currie, J. & Hyson, R. 1999 Is the impact of health shocks cushioned by socioeconomic status? The case of low birthweight. *American Economic Review* **89** (2), 245–250.

Currie, J. & Madrian, B. C. 1999 Health, health insurance and the labor market. In: *Handbook of Labor Economics* (O. C. Ashenfelter & D. Card, eds). Elsevier Science, Amsterdam, Netherlands, pp. 3309–3416.

Danert, K. & Hutton, G. 2020 Shining the spotlight on household investments for water, sanitation and hygiene (WASH): let us talk about HI and the three 'T's. *Journal of Water, Sanitation and Hygiene for Development* **10** (1), 1–4.

Gautam, S. 2018 Household (under) adoption of sanitation: Externalities and borrowing constraints. Unpublished Manuscript.

Guiteras, R., Levinsohn, J. & Mobarak, A. M. 2015 Sanitation subsidies. Encouraging sanitation investment in the developing world: a cluster-randomized trial. *Science* **348** (6237), 903–906.

Guiteras, R., Levinsohn, J. & Mobarak, A. M. 2019 *Demand Estimation with Strategic Complementarities: Sanitation in Bangladesh*. Centre for Economic Policy Research (CEPR) Discussion Paper DP13498. Available from: https://cepr.org/active/publications/discussion_papers/dp.php?dpno=13498 (accessed 17 August 2020).

Hathi, P., Haque, S., Pant, L., Coffey, D. & Spears, D. 2017 Place and child health: the interaction of population density and sanitation in developing countries. *Demography* **54** (1), 337–360.

Heckman, J., Ichimura, H. & Todd, P. 1997 Matching as an econometric evaluation estimator: evidence from evaluating a job training programme. *Review of Economic Studies* **64** (4), 605–654.

Hutton, G. & Haller, L. 2004 *Evaluation of the Non-Health Costs and Benefits of Water and Sanitation Improvements at Global Level*. Report WHO/SDE/WSH/04.04, Water Sanitation and Health Programme, World Health Organization, Geneva, Switzerland.

Hutton, G. & Varughese, M. 2016 *The Costs of Meeting the 2030 Sustainable Development Goal Targets on Drinking Water, Sanitation, and Hygiene*. Water and Sanitation Program Technical Paper, World Bank, Washington, DC, USA. Available from: https://www.worldbank.org/en/topic/water/publication/the-costs-of-meeting-the-2030-sustainable-development-goal-targets-on-drinking-water-sanitation-and-hygiene (accessed 17 August 2020).

Jappelli, T. & Pistaferri, L. 2010 The consumption response to income changes. *Annual Review of Economics* **2**, 479–506.

Kresch, E., Lipscomb, M. & Schechter, L. 2020 Externalities and spillovers from sanitation and waste management in urban and rural neighborhoods. *Applied Economic Perspectives and Policy* **42** (3), 395–420.

Leuven, E. & Sianesi, B. 2014 *Psmatch2: Stata Module to Perform Full Mahalanobis and Propensity Score Matching, Common Support Graphing, and Covariate Imbalance Testing*. Statistical Software Components.

Mara, D. & Evans, B. 2018 The sanitation and hygiene targets of the sustainable development goals: scope and challenges. *Journal of Water, Sanitation and Hygiene for Development* **8** (1), 1–16.

Novotný, J., Hasman, J. & Lepič, M. 2018 Contextual factors and motivations affecting rural community sanitation in low- and middle-income countries: a systematic review. *International Journal of Hygiene and Environmental Health* **221** (2), 121–133.

OECD 2011 *Benefits of Investing in Water and Sanitation: An OECD Perspective*. OECD Publishing, Organisation for Economic Co-operation and Development, Paris, France.

Patil, S. R., Arnold, B. F., Salvatore, A. L., Briceño, B., Ganguly, S., Colford, J. M. & Gertler, P. J. 2014 The effect of India's total sanitation campaign on defecation behaviors and child health in rural Madhya Pradesh: a cluster randomized controlled trial. *PLOS Medicine* **11** (8), e1001709.

Pearson, J. & McPhedran, K. 2008 A literature review of the non-health impacts of sanitation. *Waterlines* **27** (1), 48–61.

Pickering, A. J., Djebbari, H., Lopez, C., Coulibaly, M. & Alzua, M. L. 2015 Effect of a community-led sanitation intervention on child diarrhoea and child growth in rural Mali: a cluster-randomised controlled trial. *The Lancet Global Health* **3** (11), e701–e711.

Prüss-Ustün, A., Bartram, J., Clasen, T., Colford, J. M., Cumming, O., Curtis, V., Bonjour, S., Dangour, A. D., De France, J., Fewtrell, L., Freeman, M. C., Gordon, B., Hunter, P. R., Johnston, R. B., Mathers, C., Mäusezahl, D., Medlicott, K., Neira, M., Stocks, M., Wolf, J. & Cairncross, S. 2014 Burden of disease from inadequate water, sanitation and hygiene in low- and middle-income settings: a retrospective analysis of data from 145 countries. *Tropical Medicine and International Health* **19** (8), 894–905.

Rosenbaum, P. & Rubin, D. 1983 The central role of the propensity score in observational studies for causal effects. *Biometrika* **70** (1), 41–55.

Sclar, G. D., Garn, J. V., Penakalapati, G., Alexander, K. T., Krauss, J., Freeman, M. C., Boisson, S., Medlicott, K. O. & Clasen, T. 2017 Effects of sanitation on cognitive development and school absence: a systematic review. *International Journal of Hygiene and Environmental Health* **220** (6), 917–927.

Shakya, H., Christakis, N. & Fowler, J. 2015 Social network predictors of latrine ownership. *Social Science & Medicine* **125**, 129–138.

Spears, D. & Lamba, S. 2016 Effects of early-life exposure to rural sanitation on childhood cognitive skills: evidence from India's Total Sanitation campaign. *Journal of Human Resources* **15** (2), 298–327.

Stopnitzky, Y. 2017 No toilet no bride? intrahousehold bargaining in male-skewed marriage markets in India. *Journal of Development Economics* **127**, 269–282.

Strauss, J. & Thomas, D. 1998 Health, nutrition, and economic development. *Journal of Economic Literature* **36** (2), 766–817.

UN 2015 *Transforming Our World: The 2030 Agenda for Sustainable Development*. United Nations, New York, USA.

Available from: https://sustainabledevelopment.un.org/post2015/transformingourworld/publication (accessed 17 August 2020).

UN 2016 *Clean Water and Sanitation: Why It Matters.* Sustainable Development Goals Brief, United Nations, New York, USA. Available from: https://www.un.org/sustainabledevelopment/wp-content/uploads/2016/08/6_Why-It-Matters-2020.pdf (accessed 17 August 2020).

UNICEF and WHO 2019 *Progress on Household Drinking Water, Sanitation and Hygiene 2000–2017. Special Focus on Inequalities.* United Nations Children's Fund and World Health Organization, New York, USA. Available from: https://washdata.org/sites/default/files/documents/reports/2019-07/jmp-2019-wash-households.pdf (accessed 17 August 2020).

WHO 2019 *National Systems to Support Drinking-Water, Sanitation and Hygiene: Global Status Report 2019. UN-Water Global Analysis and Assessment of Sanitation and Drinking-Water (GLAAS) 2019 Report.* World Health Organization, Geneva, Switzerland. Available from: https://www.who.int/water_sanitation_health/publications/glaas-report-2019/en/ (accessed 17 August 2020).

WSP 2004 *Who Buys Latrines, Where and Why?* Water and Sanitation Program Field Note, World Bank, Washington, DC, USA.

First received 6 May 2020; accepted in revised form 20 August 2020. Available online 6 October 2020

Research Paper

Understanding the costs of urban sanitation: towards a standard costing model

Tristano Sainati, Fiona Zakaria, Giorgio Locatelli, P. Andrew Sleigh and Barbara Evans

ABSTRACT

There is a dearth of reliable cost data for urban sanitation. In the absence of high-quality global data, the full cost of sustainable implementation of urban sanitation remains uncertain. This paper proposes an approach for developing bespoke parametric cost estimation models for easy and reliable estimation of the costs of alternative sanitation technologies in a range of geographical contexts. A key requirement for the development of these models is the establishment of a large database of empirical information on the current costs of sanitation systems. Such a database does not currently exist. Two foundational tools are proposed. Firstly, a standard metric for reporting the costs of urban sanitation systems, total annualised cost per household. Secondly, a standardised approach to the collection of empirical cost data, the Novel Ball-Park Reporting Approach (NBPRA). Data from the NBPRA are presented for 87 individual sanitation components from 25 cities in 10 countries. Broad cost ranges for different archetypal systems have been estimated; these currently have high levels of uncertainty. Further work is proposed to collect additional data, build up the global database, and develop parametric cost estimation models with higher reliability.

Key words | benchmarking, cost estimation, costing standards, urban sanitation

Tristano Sainati (corresponding author)
Fiona Zakaria
Giorgio Locatelli
P. Andrew Sleigh
Barbara Evans
Water@leeds, University of Leeds,
Leeds,
UK
E-mail: *t.sainati@leeds.ac.uk*

HIGHLIGHTS

- A bespoke parametric cost estimation approach is proposed for reliable estimation of urban sanitation costs.
- A standard metric for reporting the costs of urban sanitation systems; total annualised cost per household (TACH) is presented.
- Broad cost ranges for different archetypal systems have been estimated, based on 87 individual sanitation components from 25 cities in 10 countries.

INTRODUCTION

More than 1 million people per day must gain and maintain access to safely managed sanitation to meet the sanitation target of the sustainable development goals (Mara & Evans 2018). A significant majority of this 'new' sanitation investment will be made in urban or urbanising areas. While standalone household and community sanitation services will continue to be important in rural areas (in both rich and poor countries) in denser urban areas, toilets that

doi: 10.2166/washdev.2020.093

provide sanitation at home must be planned and operated as part of a professionalised service – which transports excreta either through pipes or road-based networks for management away from the home.

A key challenge for SDG 6.2 arises from the dearth of reliable and comparable benchmark estimates of the unit costs of these 'networked' services (Daudey 2017). Compared to other infrastructure services, costing in sanitation is characterised by ambiguity concerning both terminology and costing standards. Other sectors tend to employ familiar, readily understood, summary indicators for the global comparison of costs (Merrow 2011; Ansar *et al.* 2014; Locatelli *et al.* 2017). For example, the concept of the cost per kilowatt-hour (kWh) of energy is well established (De Roo & Parsons 2011; Geissmann & Ponta 2017; Lai & McCulloch 2017) and its utility in comparing the costs of services can be easily understood even by a layperson. By contrast, there is no agreed metric for comparison of the costs of sanitation delivery (Daudey 2017).

This paper reports early findings from the Cost and Climate for Urban Sanitation (CACTUS) project which aims to fill some of this gap. We introduce a proposed standard cost metric for urban sanitation, Total Annualised Cost per Household (TACH) and per Capita (TACC), and propose a strategy for developing a parametric method that would enable the estimation of TACH/TACC for new sanitation systems around the world. We also describe a Novel Ball-Park Reporting Approach (NPBRA) which is aiding in the establishment of a global database of empirical benchmark costs as a basis for this future parametric cost estimation tool. Indicative early results from the global database of NPBRA data are presented.

THE CHALLENGE OF COSTING IN URBAN SANITATION

In a recent systematic review, Daudey (2017) observes a dearth of reliable studies reporting clearly on the full costs associated with the delivery of sanitation services worldwide. While there has been notable and important work done on estimating benefit–cost ratios in sanitation, see for example, Whittington *et al.* (2012), Hutton 2013 & Cronin *et al.* (2014), these rely on a very small number of empirical data and limited modelling to derive the cost estimates on which they are based.

One approach to estimating the cost of services is to use the price as a proxy. Unfortunately, price is a poor proxy for the cost of a sustainable sanitation service for two reasons. Firstly, many sanitation systems are incomplete or badly operated often due to financial, institutional, and societal failures (Evans 1995; Ika *et al.* 2012; International Water Association 2014). Secondly, since sanitation is a public good, willingness to pay is often lower than the economic value of full-service delivery (Ruiters & Matji 2017). For this latter reason, sanitation systems are often supported by public subsidies, either formally through regulated fees or 'informally' through systematic underinvestment in maintenance (Evans *et al.* 2009; Perard 2018).

The best-known sanitation costing studies are from the World Health Organization (Hutton & Bartram 2008), the IRC WASHCost Initiative (McIntyre *et al.* 2014), and the World Bank Economics of Sanitation Initiative, or ESI. These all emphasise the importance of reporting the full lifecycle cost of sanitation systems, including costs incurred to design, operate, and maintain them (Fonseca *et al.* 2011; McIntyre *et al.* 2014). Lifecycle costing implicitly assumes a full-costing approach (Burr & Fonseca 2011; Evans & Mara 2011; Daudey 2017), including the direct cost of generating the sanitation service and overheads. In reality, there is a significant ambiguity around the methods used to generate unit cost estimates and particularly the extent to which operations and maintenance costs are reflected in many costing studies (Daudey 2017).

Hutton & Bartram (2008) and World Health Organization (2012) deploy an approach known as **ball-park estimation.** Ball-park estimation is commonly used in infrastructure planning for conceptualisation or pre-feasibility studies (Mislick & Nussbaum 2015) where some level of uncertainty is acceptable; construction projects typically accommodate an uncertainty at this point of around 30% (AACE 2019). Estimations are usually based on the statistical manipulation of historical data in private databases (e.g. the internal information systems of Engineering Procurement Construction – EPC companies) or academic literature (OECD 2007; Loutatidou *et al.* 2014; Hughes *et al.* 2015).

Ball-park estimates from a number of locations both reported by governments and cited in the literature, particularly the ESI, form the basis for the indicative unit costs estimates in the WHO benefit–cost analysis (Hutton & Haller 2004; Hutton & Bartram 2008; World Health Organization 2012). The economics of sanitation initiative has collected cost estimates based on reporting by government agencies, and household surveys, for a number of countries, see for example, Hutton *et al.* (2014). WASHCost also aimed to generate ball-park cost estimates based on empirical data collection rather than relying on literature. WASHCost reported costs data from water and sanitation projects in rural and peri-urban/small town areas in the state of Andhra Pradesh (India), Burkina Faso, Ghana, and Mozambique. Lifecycle cost estimates were developed for seven types of latrine (Burr & Fonseca 2012).

While WASHCost highlighted the need for standard costing approaches and has played an important role in increasing focus on the significant operational costs of sanitation, there is little evidence that the approach has been used to generate up to date benchmark cost estimates for *urban* sanitation. Neither is WASHCost referenced in the wider infrastructure finance literature; this may be because it uses non-standard cost categories, for example, the term Capital Maintenance Expenditure (CapManEx) which is neither a common accounting term (Roman *et al.* 2012) nor employed in other infrastructure sectors (De Schepper *et al.* 2015; Gatti 2018). The economics of sanitation initiative did generate useful overall cost estimates for some technologies in urban areas (as well as in rural areas) which provide a strong starting point for the current work.

An alternative to ball-park estimation, **analytical estimation** works on the basis of a known or estimated amount of inputs (material, labour, equipment) with known unit costs; the full costs can then be obtained by multiplying one by the other. The CLARA project (Müllegger *et al.* 2014; Ketema & Langergraber 2016) is an example of this approach building up theoretical costs from first principles based on standard bills of quantities (Müllegger *et al.* 2014).

CLARA used empirical data from Burkina Faso, Ethiopia, Kenya, Morocco, and South Africa to develop a simplified analytical costing tool (Müllegger *et al.* 2014; Ketema & Langergraber 2016). The tool can be used to generate cost estimates for sanitation based on local conditions and the Net Present Value of the investment (Ketema *et al.* 2015). The World Bank has used a similar approach in its faecal sludge management (FSM) diagnostic tools (Ross *et al.* 2016).

Analytical estimation works when sufficiently detailed information is available for both quantities and unitary prices of all factors of production (Mislick & Nussbaum 2015), including capital (e.g. land and buildings, and equipment) and labour (Kim *et al.* 2004; Barakchi *et al.* 2017). However, the project management literature emphasises the extent to which analytic estimation can be unreliable particularly for large engineering projects. A number of phenomena tend to lead to over-optimistic estimations (Locatelli 2018). These include optimism bias (Flyvbjerg 2008), strategic misinterpretation (Flyvbjerg 2008), corruption (Locatelli *et al.* 2017), selection biases and winner course phenomenon (Eliasson & Fosgerau 2013), and distortions in tendering (Love *et al.* 2016, 2019). For this reason, some scholars suggest using ball-park estimation for large and complex infrastructure projects, even when detailed analytic data are available (Flyvbjerg 2008; Merrow 2011).

An alternative approach is **parametric cost estimation** (also known as factorial or semi-analytic) costing. This approach recognises that there are critical drivers (parameters) that determine the cost of a service or infrastructure in a given context. The traditional approach to developing parametric cost estimation is to interrogate a large dataset of empirical cost data, employing heuristic formulas to determine the relationship between various parameters and cost values (Mislick & Nussbaum 2015; OECD 2007). In sanitation, some attempts at parametric cost estimation have been made to estimate the nominal costs of different sanitation technologies (Sinnatamby *et al.* 1986; Mara 1996; Mara & Guimarães 1999; Evans & Mara 2011; Eggimann *et al.* 2016a, 2015; Crocker *et al.* 2017; Loetscher & Keller 2018).

Parameters that primarily affect the costs of sanitation systems have been proposed and evaluated by several scholars; these include population densities, size and degree of centralisation (Eggimann *et al.* 2015, 2016b; Eggimann 2016), economies of scale (Hernández-Chover *et al.* 2018), institutional and managerial context (Ika & Donnelly 2017), and other contextual factors such as the technology, labour cost, population density, and topography (Whittington *et al.* 2012; Daudey 2017). Unfortunately, there has been no systematic attempt to evaluate the relationship between these parameters and cost based on empirical data at the required scale. The results remain specific to isolated contexts or are based on models which have not been empirically validated.

There remain, therefore, significant areas of further research in the arena of sanitation costing, including the need for (Daudey 2017; Eggimann *et al.* 2017):

- Further empirical research to generate a larger database from which both ball-park and parametric estimates could be derived;
- Improvements in the consistency of reporting of cost estimations;
- Further research to assess the full-lifecycle cost of sanitation services, and the differentiation between alternative sanitation technologies, stages of the sanitation value chain and geographical areas; and
- More clarity and transparency concerning the sources of empirical data used in current costing approaches [....]

THE STANDARD COST METRIC FOR URBAN SANITATION; TACH AND TACC

Responding to the call for greater consistency and for generalisable information about the cost of sanitation, we propose a plausible metric which is comparable across geographies and technologies and also understandable both to professionals in the WASH sector and to municipal managers who make decisions regarding sanitation investments. A useful corollary is the concept of costs per kWh of energy which is comprehensible both to governments, who may have to finance the initial capital investment, and households, who ultimately pay the bills.

For urban sanitation, we propose two cost indicators to express the cost of any sanitation system: TACH and total annualised cost per capita (TACC). TACH/TACC takes into account full lifecycle costs which are annualised and expressed on a per-household or per-user basis.

The key question in any infrastructure investment is the total level of financial liability that the planned investment is likely to generate compared to total sources of income (which are limited to tariffs, tax revenue, and transfers from central government or development partners). The liability associated with the proposed investment includes both the costs of capital (the investment, plus the financing costs if any) plus the operational costs over the full lifetime of the project, the lifecycle cost reported by Fonseca *et al.* (2011). Operational costs for some systems are often excluded from a discussion of the relative costs of different systems – possibly since they are assumed to be covered by households (fees charged by operators who empty onsite pit latrines and tanks are an example). However, we argue that information on the full-cost liability is essential to enable municipal decision-makers to incorporate sanitation investments into municipal budgeting. Full-cost information facilitates the development of plans to sustainably cover all the costs that are associated with the planned investment; in the absence of this full cost information, operational costs are often underestimated or omitted, with resultant underfunding and system failure.

However, the total lifecycle cost is effectively meaningless at the local government level, where budgets are managed on an annual basis. For this reason, it makes the most sense to convert this to an annual liability, covering debt service for capital investment plus the annual operational costs and periodic maintenance requirements.

To enable comparison across various scales of systems, the total number of people directly using the service is used as the denominator. The choice of household or capita for the denominator is challenging – each has merit. Per capita costs are perhaps most useful when looking at national level macro-economic performance, or when comparing expenditure on sanitation against that on say the police force, but some local governments rely on property tax for their revenue, and water and sewerage utilities bill on a household basis. For many decision-makers therefore using the household as a denominator makes the most sense.

A PROPOSED APPROACH TO COST ESTIMATION

In selecting an approach to use in estimating the TACH or TACC for any given system in a given place, there is a trade-off between precision, reliability, and level of detail on the one hand, and availability of data and resources on the other. The focus here is on supporting local decision-makers to make credible plans quickly and cheaply, selecting the best technical options in a given context (Mitlin 2015). This requires costing data that are reliable, comparable, and easy to make locally relevant (Kalbermatten *et al.* 1982). We rejected analytic estimation on the basis that it has very high information requirement, and consequently high cost, yet remains prone to bias (Flyvbjerg *et al.* 2018; Love *et al.* 2019).

Instead, the proposed approach to cost estimation in the CACTUS project is to develop a parametric estimation tool that could generate plausible cost estimates for a range of sanitation technologies and systems in any given context based on the known values of these key parameters.

From earlier reviews, several key contextual parameters have been identified as candidates in determining the relative cost of different sanitation systems. These are scale (size of city and size of the sanitation system); population density; topography; and geographical location (as a proxy for relative costs of inputs such as labour, materials, fuel) (Whittington *et al.* 2012; Eggimann *et al.* 2015, 2016b; Eggimann 2016; Daudey 2017; Ika & Donnelly 2017; Hernández-Chover *et al.* 2018).

Having identified a set of candidate parameters for cost estimation purposes, the next task is to assemble a reliable database which could be used to develop the heuristics needed for the estimation approach to work. As already mentioned, data are scarce. Consequently, an important element of our work is the collection of new data on which to base our estimations.

Some specific challenges arise, however, in the collection and organisation of urban sanitation cost data. Urban sanitation services are provided by a compound of capital-intensive infrastructure, operational management, and ongoing interventions, including capacity building, advocacy, and other promotional activities all of which happen along the sanitation 'value chain'. With the exception of fully operational sewered networks, very few urban sanitation chains are operated by a single operator 'end to end', and many sanitation systems are not fully functional. Cost data reported by operators are often therefore only a partial representation of the real costs of delivery of safely managed sanitation. To address these challenges, we set out to build a database of cost information that could be used to estimate the full costs of service delivery for the entire value chain. To achieve this, we developed an NBPRA for sanitation.

NBPRA FOR SANITATION

The NBPRA facilitates the collection and collation of data in a consistent format from operators who are delivering ongoing sanitation services. Cases are selected from a broad range of contexts (covering a wide range of parameters) and a range of different sanitation technologies and approaches.

The approach is standardised and based on four main pillars, namely (1) technological homogeneity, (2) acceptable service, (3) basic costing assumption, and (4) the reference business model. Cost data are then normalised to comparable currency equivalent values, annualised, and divided by the household/people served, so they can be reported as TACH/TACC.

Technological homogeneity

The NBPRA is exclusively concerned with a subset of 'safely managed' sanitation as defined by the Joint Monitoring Programme for Water Supply, Sanitation, and Hygiene (JMP), namely systems which guarantee that households use 'an improved type of sanitation facility that is not shared with other households and the excreta .... [is] transported and treated off-site.' (UNICEF & WHO 2019). In reality, many systems that use appropriate technologies are incomplete, for example, faecal sludge which is transported but then dumped in the environment (Peal *et al.* 2020). The NBPRA is designed to accommodate empirical data collected from projects and programs that only deliver partial services (toilets connected to septic tanks, for example, or an emptying service). Cost estimates generated for these partial sanitation systems can be added together to generate cost estimates for complete sanitation systems. To enable this summing, partial service types need to be clustered into meaningful technology categories which are internally consistent, externally comparable in technological terms, and amenable to interpretation by both sanitation experts and the wider community of decision-makers.

The NBPRA, therefore, uses a set of standard 'components' definitions, each of which maps to a single 'element' of the sanitation value chain (containment, emptying, transport, treatment, etc.). The 'components' were selected to reflect the most common technological approaches to deliver networked (urban) sanitation. Components are based on Tilley *et al.* (2014) but have been bundled somewhat to reduce complexity and facilitate a meaningful comparison on a cost basis. We use general terminology, recognising that specific local terms are often used but that these often have different meanings in different locations (for example, despite the existence of a globally recognised technical definition, the terms 'septic tank' and 'pit latrine' often have a different precise connotation in different places and we therefore avoid their use here).

For wastewater systems, we assume that the main technological driver of the cost relates to the nature of the sewer network specifically;

- Conventional versus simplified networks;
- Combined versus separate networks;
- Pumped versus gravity-dependent networks.

Combining these options leaves a set of eight component types for emptying and transport in wastewater based systems (see Figure 1).

For systems which rely on pits or tanks (often referred to as 'onsite' or FSM systems) and container-based systems, we differentiate four main types of component at the containment (household) end of the sanitation value chain (sealed tank with an infiltration structure, sealed tank without an infiltration structure, infiltrating pit, and container for container-based systems). Moving excreta from these systems can either take place in two steps (emptying plus transport) or in a single operation (typically where a single truck is used to empty the tank and transport the contents away to treatment). In either case, we make a distinction between manual powered systems and externally powered systems (Figure 1).

For treatment, we have distinguished between aerated and anaerobic systems for both wastewater and faecal sludge, and in the case of wastewater, we further distinguish between passive and powered aeration. A summary of the 27 final component types is shown in Figure 1.

Acceptable service

Many otherwise 'appropriate' sanitation systems fail due to underfunding, for example, sewer systems that leak due to low maintenance (Peal *et al.* 2020). These funding gaps can be identified and filled, when sufficient empirical data are available, through modelling, to give a clear idea of the full costs of the system if it were to be properly operated and maintained.

Clearly, there are variations in the 'level of service' provided by different sanitation systems – private, shared, community, and public toilets offer a wide range of levels of amenity to users (Evans *et al.* 2017); different treatment options will perform differently in terms of efficiency (Choudhury & Saha 2017; Lutterbeck *et al.* 2017), environmental impact, embodied energy waste recovery (Cornejo *et al.* 2013), resilience (Luh *et al.* 2017), and sustainability (Zurbrügg *et al.* 2014).

These aspects of the sanitation system performance are important. However, rather than internalising them resulting in greater complexity and less transparency, the NBPRA explicitly excludes them. The aim is to generate reliable comprehensive cost estimates for systems that deliver 'safely managed sanitation' to which decision-makers may choose to add other performance criteria when making real-world technology selections.

Basic costing assumption

The NBPRA is a full-costing approach (Arnaboldi *et al.* 2014), generating cost data which can subsequently be used for capital budgeting. and focused on the financial cost (Roman *et al.* 2012; Arnaboldi *et al.* 2014). Data are reported in terms of the 'industrial cost', i.e. the full cost of sanitation service consisting of both the direct and indirect cost of building and operating sanitation infrastructure, plus administrative charges, the cost of financing, and taxation.

The costs associated with any component are categorised using a standard proforma to facilitate both data collection (acting as a checklist for enumerators), data verification, and modelling of incomplete data (for example, where an operator cannot or will not report the cost of land, or where taxes and fees are not paid). Cost categories used within the NBPRA are shown in Table 1.

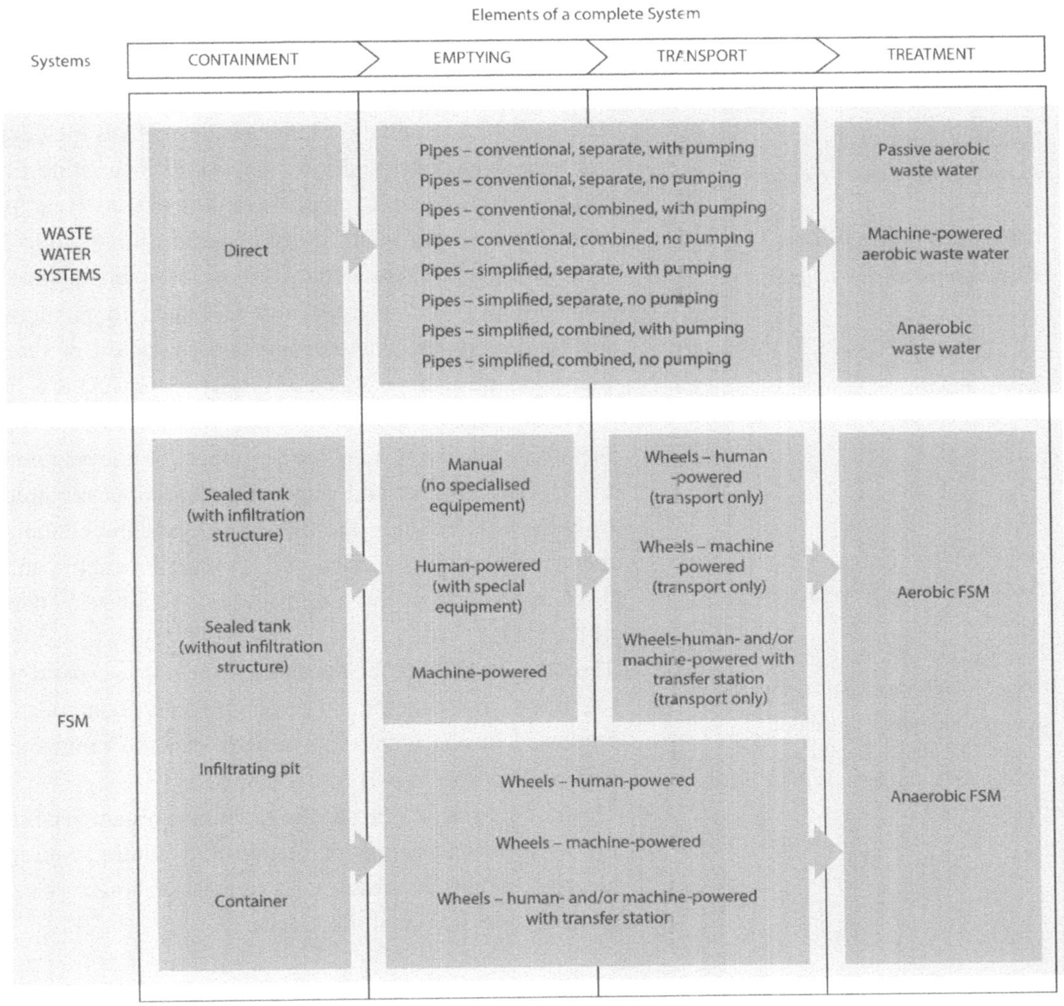

Figure 1 | Components of a sanitation system for use in CACTUS cost estimation, distinguishing between systems that predominantly rely on pipes to transport excreta in the form of wastewater, and all other systems which are often loosely categorised as faecal sludge management (FSM)-based systems.

Reference business model

Sanitation services can be provided via a wide range of business models with varying degrees of centralisation, and aggregation along the value chain. Because of this complexity, there is a challenge in reporting on the basis of a standardised unit of analysis consistent in regard to (1) number and types of transactions, (2) ownership structure, and (3) organisational structure (Williamson 1979, 1981). Variations in these characteristics introduce distortions in the reporting of empirical data jeopardising their reliability and the ability to create a consistent sample of cost benchmarks.

To overcome this challenge, the NBPRA assumes:

1. There is a 'virtual' organisation responsible for operating the entire sanitation system herein called the unique operator;
2. The unique operator owns all the sanitation infrastructure. The operator finances the sanitation infrastructure and borrows all or part of the capital on the financial market at market value;
3. The unique operator buys (at market value) all the factors of production that are necessary to generate sanitation services across the entire sanitation value chain;
4. The unique operator hires and manages all necessary labour (direct and indirect) for generating the sanitation services across the entire sanitation value chain.

Table 1 | Standard CACTUS framework for reporting costs of urban sanitation

Cost voice	Examples
Capital costs – CAPEX	
Land	*Land purchase*
Infrastructure and building equipment	*Buildings, fixed plant purchase*
Equipment	*Plant and vehicle purchase*
Staff development	*Vaccinations, one off training*
Major and extraordinary repairs	
Other CAPEX	
Administration charges	
Financing	
Taxes	
Operational costs – OPEX	
Land	*Land rental*
Infrastructure and building equipment	*Leasing costs for buildings and fixed plant*
Equipment	*Leasing costs, operational costs*
Staffing	*Salaries, pensions, health insurance*
Consumables - Utilities - Fuel - Chemicals - Services - Other	Services includes: Consulting/ advisory, Legal, Insurance, Regular Maintenance, other
Other OPEX	
Administration charges	
Financing	
Taxes	

These principles introduce an archetypical business model that never materialises in real life but enables the cost data that are collected to be used to generate costs for a range of complete sanitation systems which are applicable irrespective of the actual business model context.

A key outcome of this approach is that revenue streams and transfers between operators of different elements of the sanitation system (for example, payment by a truck operator when dumping at a treatment plant) are excluded. They are exogenous to the actual costs of implementing the safely managed sanitation service and do not correspond to any industrial activity for the generation of the sanitation service.

As an example, taking a typical FSM system based on sealed tanks at household level, the following 'industrial' costs are included:

1. The cost of the sanitation infrastructure which includes the household toilet and the associated substructure (in this case a sealed tank), emptying and conveyance equipment (e.g. vacuum tank), transfer station (if any), and a treatment plant. These include the cost of financing infrastructure (e.g. the interests and fees paid to financial institutions for the construction and operation of sanitation infrastructure). This cost is part of the CAPEX;
2. The direct cost of operating the sanitation infrastructure, including the input supplies (e.g. fuel for the vacuum truck), consumables and wages (e.g. the labour required to clean and maintain the toilet, emptying the containment technologies, drive the vacuum truck, and discharge and treat the faecal sludge). This cost is part of the OPEX;
3. The cost of indirect facilities (e.g. office space, parking lot, etc.) and equipment (e.g. phones, computers) employed by the unique operator to run the sanitation system. This cost is part of the CAPEX;
4. The indirect cost of running the sanitation system, including annual expense on marketing and administration, licenses, public concession for the sanitation service, etc. This cost is part of the OPEX.

Normalisation process

In CACTUS, empirical data are collected from a wide variety of locations, for sanitation systems that have varying life expectancy and technical characteristics. To facilitate the parametric cost estimation in the future, these costs are converted to a standard reporting year and currency and annualised, to generate a comparable annual cost liability in each case.

Currency conversion generates costs in terms of International \$ 2018 (Int\$ 2018), based on the Consumer Price Index using Purchasing Power Parity (Rao 2001; Lakner *et al.* 2018). Conversion factors are based on the World Bank Database (World Bank 2020).

Annualisation is applied in line with Stewart *et al.* (1995). The cost is expressed as the Equivalent Annual

Table 2 | Cost per household for sanitation components based on empirical and secondary data sources

			Total annualised cost per household (TACH) (Int$ 2018)					Annualised CAPEX per household (Int$ 2018)					Annual OPEX per household (Int$ 2018)				
	Element	**Component**	**n**	**Median**	**Mean**	**Min**	**Max**	**n**	**Median**	**Mean**	**Min**	**Max**	**n**	**Median**	**Mean**	**Min**	**Max**
Wastewater	Containment	Direct	1	**362**	362	362	362	6	**94**	103	34	233	1	**130**	130	130	130
	Emptying and transport	Pipes – conventional, separate, with pumping	5	**195**	395	107	832	5	**164**	296	60	668	5	**61**	99	31	195
		Pipes – conventional, separate, no pumping	–	No data				1	**704**	704	704	704	–	No data			
		Pipes – conventional, combined, with pumping	7	**287**	294	93	515	7	265	263	75	473	7	**37**	31	17	42
		Pipes – conventional, combined, no pumping	–	No data				–	No data				–	No data			
		Pipes – simplified, separate, with pumping	–					–					–				
		Pipes – simplified, separate, no pumping	–					–					–				
		Pipes – simplified, combined, with pumping	–					–					–				
		Pipes – simplified, combined, no pumping	–					–					–				
	Treatment	Passive aerobic waste water	6	**28**	53	11	148	7	**8**	64	1	282	7	**10**	21	1	80
		Machine-powered aerobic waste water	15	**134**	159	58	315	15	**99**	106	21	239	15	**54**	54	2	131
		Anaerobic waste water	–	No data				–	No data				–	No data			

(continued)

Table 2 | continued

	Element	Component	Total annualised cost per household (TACH) (Int$ 2018)					Annualised CAPEX per household (Int$ 2018)					Annual OPEX per household (Int$ 2018)				
			n	Median	Mean	Min	Max	n	Median	Mean	Min	Max	n	Median	Mean	Min	Max
Faecal Sludge Management (FSM)	Containment	Sealed tank (with infiltration structure)	–	No data				–	No data				–	No data			
		Sealed tank (without infiltration structure)	3	**87**	84	67	97	8	**39**	52	8	165	3	**28**	34	26	49
		Infiltrating pit	4	**63**	95	30	223	4	**63**	95	30	223	4[a]	**0**	0	0	0
		Container	2	**115**	115	106	123	2	**47**	47	34	61	2	**67**	67	62	72
	Emptying	Manual (no specialised equipment)	2	**82**	82	48	116	2[b]	**4**	4	0	9	2	**78**	78	48	107
		Human-powered with specialised equipment	2	**25**	25	23	27	2	**0**	0	0	0	2	**25**	25	22	27
		Machine powered	–	No data				–	No data				–	No data			
	Emptying and transport	Wheels – human-powered	2	**136**	136	109	163	2	**2**	2	1	3	2	**134**	134	108	160
		Wheels – machine-powered	11	**28**	29	3	83	11	**5**	7	1	32	13	**13**	19	0	51
		Wheels – human- and/or machine-powered with transfer station	1	**83**	83	83	83	1	**1**	1	1	1	1	**82**	82	82	82
	Transport	Wheels – human-powered (transport only)	–	No data				–	No data				–	No data			
		Wheels – machine-powered (transport only)	1	**26**	26	26	26	1	**8**	8	8	8	1	**17**	17	17	17
		Wheels – human- and/or machine-powered with transfer station (transport only)	1	**1**	1	1	1	1	**1**	1	1	1	1	**1**	1	1	1
	Treatment	Aerobic FSM	5	**17**	36	0	103	5	**5**	9	0	25	6	**9**	23	0	94
		Anaerobic FSM	3	**44**	47	11	87	3	**21**	31	4	69	3	**17**	16	7	23

[a]For all data points for 'infiltrating pits', OPEX is reported or assumed to be zero.
[b]One data point of 'manual (no specialised equipment)' is reported to be zero.

Cost (EAC) of owning, operating, and maintaining the sanitation system (or sub-system) over its entire lifecycle. In other words, it is the annual expenditure needed to cover the servicing of capital debt required to construct and maintain the system plus the annual operational budget.

The EAC is calculated, as shown in Equation (1).

$$\text{EAC (Sanitation System or Subsystem)} = \frac{k_r \cdot \sum_{t=0}^{T} \dfrac{\text{COST}_t \cdot k_r}{(1+k_r)^t}}{1-(1+k_r)^{-T}} \quad (1)$$

where:

- COST_t = are the costs incurred during the lifecycle (i.e. T) associated with the data point considered.
- T = represents the longest lifecycle associated with the data point considered, and it is calculated in years. T corresponds to the longest lifetime of the infrastructures associated with the data point;
- kr = real interest rate, which is calculated using the following formula (Equation (2))

$$k_r = \frac{k_n - s}{1+s} \quad (2)$$

where

- kn = nominal interest rate, assumed 5% as the social discount rate.
- s = annual inflation in the country considered.

Table 3 | Synthesised cost range estimates from the CACTUS database, for example, archetypal sanitation systems with no parameterisation

	Total annualised cost per household by system element and for whole system (IntS 2018) Median (Minimum–maximum)				
Archetypal sanitation system	**Containment**	**Emptying**	**Transport**	**Treatment**	**Whole System**
Container-based sanitation, with mechanised emptying and transfer stations, with composting (aerobic treatment)	Container $n = 2$ **115** (106–123)	Wheels – human and/or machine powered with transfer station[a] $n = 1$ **83** (83–83)		Aerobic FSM $n = 5$ **17** (0–103)	**215** (189–309)
Onsite 'septic' tanks, mechanised emptying and transport with anaerobic treatment	Sealed tank without infiltration structure $n = 3$ **87** (67–97)	Wheels – machine powered $n = 11$ **28** (3–83)		Anaerobic FSM $n = 3$ **44** (11–87)	**159** (81–267)
Onsite 'septic' tanks, mechanised emptying and transport with anaerobic treatment (Africa only)	Sealed tank without infiltration structure $n = 3$ **87** (67–97)	Wheels – machine powered $n = 3$ **30** (28–83)		Anaerobic FSM[a] $n = 1$ **11** (11–11)	**128** (106–191)
Sewerage, conventional, combined, pumped, with activated sludge treatment	Direct[a] $n = 1$ **362** (362–362)	Pipes – conventional, combined, with pumping $n = 7$ **287** (93–515)		Machine-powered aerobic waste water $n = 15$ **134** (58–315)	**783** (513–1,192)

[a]$n = 1$ data have very high uncertainty and cannot be cross checked.

In calculating the EAC, most capital expenditures are considered as one-off expenses at time zero, except major and extraordinary repairs, which are considered cyclical with a period longer than one year. Operational expenditures are always expressed on an annual basis and are assumed fixed during the period *T*.

Generation of TACH and TACC estimates

TACH/TACC is calculated by dividing the EAC by the estimated number of households or people served. To aid interpretation, EAC can be calculated for CAPEX and OPEX jointly or separately. This may be useful where capital liability is carried by a different government entity than operational liability for example. The reporting unit is International $(2018) per household per year, or per person per year.

RESULTS

The NBPRA has been tested in 25 cities in 10 countries, collecting data on 87 individual sanitation components. A summary of the data collected and processed to date is shown in Table 2.

What the preliminary results mean

Typical cost liabilities of full sanitation systems of a particular type can be assessed from CACTUS data by summing indicative costs of a set of relevant components (to account for containment, emptying, transport, and treatment). To demonstrate how these can be constructed, examples of estimated cost ranges for synthetic sanitation systems are shown in Table 3. Median data are preferred to means due to the heavy skewing that can result from a single outlying data point. Further break downs of the data are possible – for example, in Table 3, we also show selected data for onsite systems in Africa. Estimates of TACH for complete systems have high levels of uncertainty at this stage in the process due to the small sample sizes and clustering of case studies. The data come from 25 cities, representing only a small sample of the conditions under which urban sanitation systems are implemented.

As might be expected, TACH for the sewerage system shown in Table 3 is significantly higher than TACH for container-based or onsite sanitation systems. However, in Figure 2, we present the data sorted by country, and showing both total costs and CAPEX/OPEX cost breakdown, all annualised on a per household basis. While sewered sanitation systems are often said to be 'more

(a)

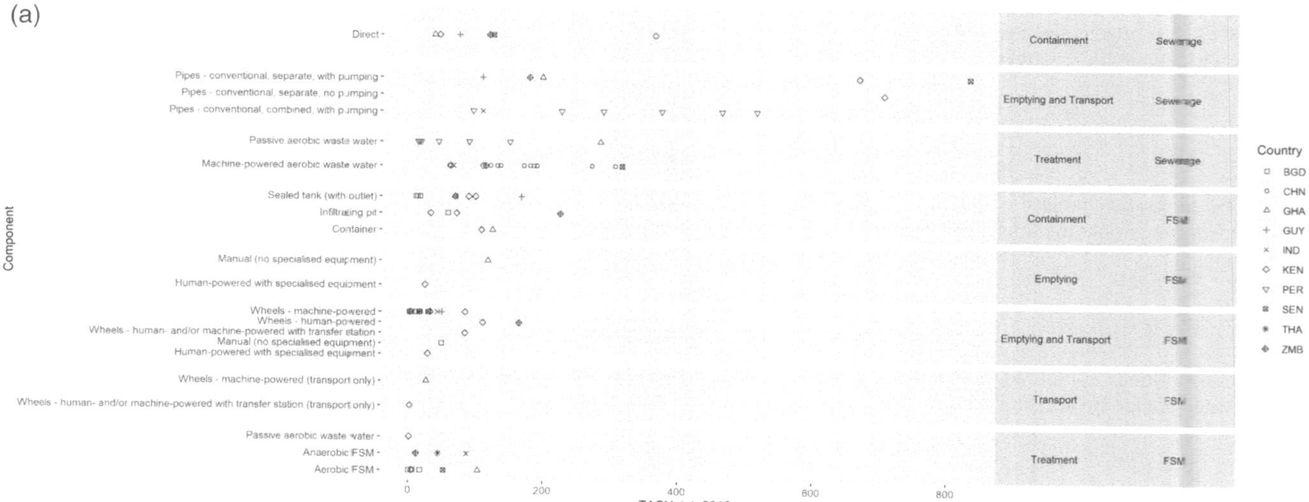

Figure 2 | (a) Total annualised cost per household for sanitation components based on empirical and secondary data in the CACTUS database ($n = 71$). (b) Annualised CAPEX per household for sanitation components based on empirical and secondary data in the CACTUS database ($n = 83$). (c) Annual OPEX per household for sanitation components based on empirical and secondary data in the CACTUS database ($n = 75$). (*continued*).

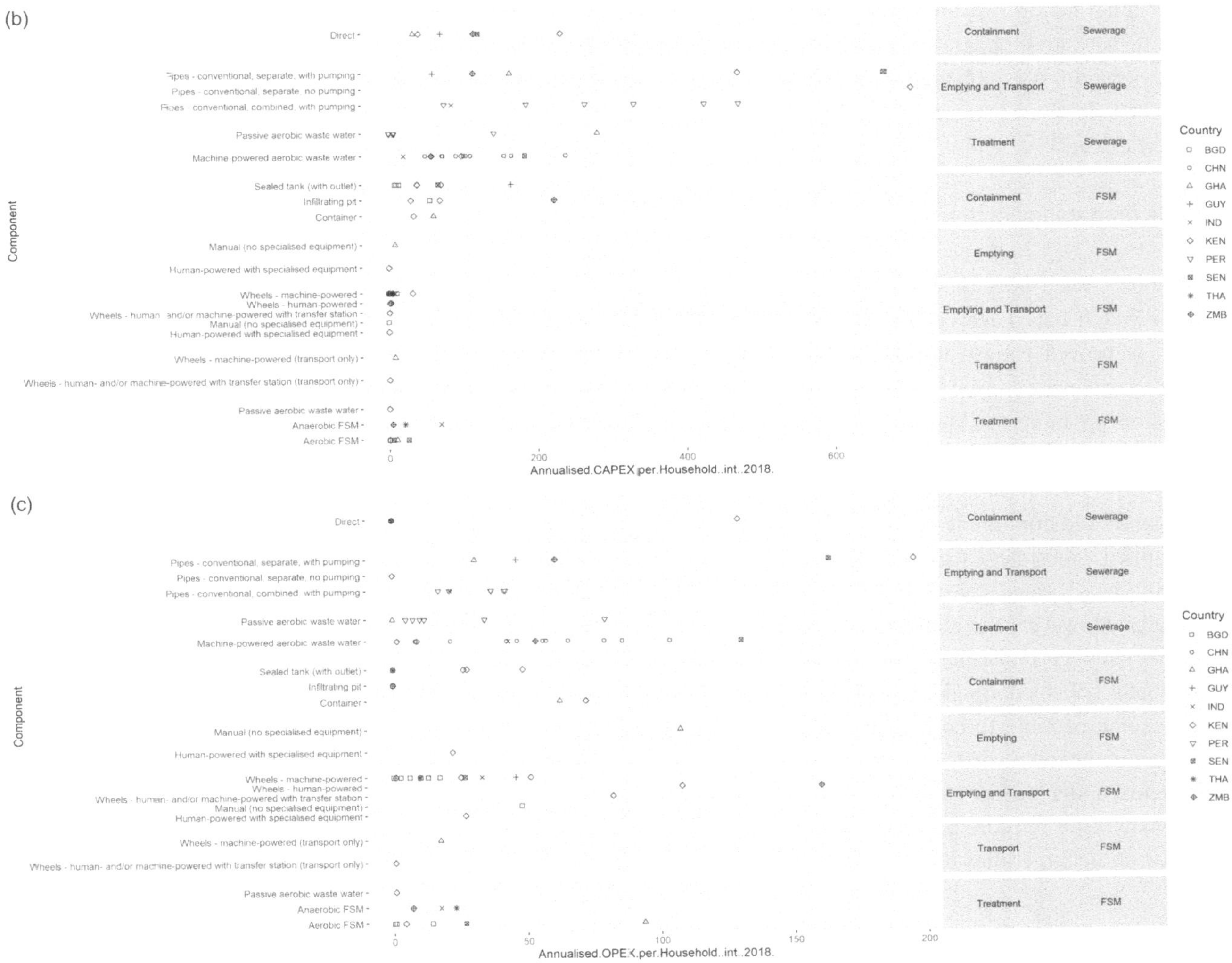

Figure 2 | (continued).

expensive' than FSM-based systems, a closer inspection of the data suggest that the situation may be more complex. For example, the operational liabilities of sewers may sometimes be lower than those for road-based transportation of faecal sludge under some conditions (see Figure 2(b)). Returning to Table 2, it is also possible to see that TACH for container-based systems is dominated by operational costs whereas onsite systems and sewers have a much higher CAPEX dependency. In all these cases, care is necessary because of the aforementioned clustering of data points.

Candidate cost drivers for parametric estimation

Figure 2 shows that for some systems, TACH clusters by country (see, for example, that the data for mechanised aerobic treatment of wastewater are higher in the China cases than in the other three countries for which data points are reported). A much larger dataset will be required to properly understand the combination of factors which drive differing cost performance in each case. While the geographical location is likely to be one driver (as it will determine for example the relative costs of materials, fuel, and labour) other factors, including the scale of the system, population density, and topography are also highly likely to drive costs variation. A significant increase in the data held in the CACTUS database is needed in order to fully interrogate the cost drivers.

Reflections on data collection

Figure 2 shows wide cost ranges for many of the components for which data have been collected. The NBPRA approach has proved robust at driving the collection of reasonably complete empirical data on costs, although many operators are unable to fully report on their cost liabilities. For older systems, there is often a lack of data on capital costs, and for newer systems, a lack of data on operational costs. In addition, many operators are not aware of certain implicit subsidies (for example, non-payment of electricity bills issued by national energy-generating organisations). However, the standardised data collection approach has shown promise in helping to drive up the quality of data that is collected. In addition, as the dataset grows, it would become possible to correct for omitted data (for example, by understanding typical cost distributions for particular systems under particular conditions). This type of correction may result in more accurate, clustered estimates of TACH for particular contexts.

CONCLUSION

Compared to other infrastructure sectors, there is a dearth of reliable, internationally comparable cost data for urban sanitation. Sanitation scholars have used both ball-park and analytical cost estimation approaches, and there have been some localised or specific efforts to generate models for parametric estimation. In the absence of reliable global data, the full costs of sustainable implementation or urban sanitation are being systematically underestimated (Flyvbjerg 2008).

This research considers the lessons learned in other sectors in addressing similar costing challenges and proposes a strategy for developing bespoke parametric cost estimation models to favour easy and reliable cost estimation, for alternative sanitation technologies, in a range of contexts. Two key requirements of such an approach are the development of standard costing metrics and the development of a large and coherent empirical dataset of sanitation technology cost estimates selected from a range of geographical context and sanitation technologies.

The main contribution to knowledge to date comprises the proposition of the TACH – TACH – and per capita – TACC, costing metrics, which are foundational to CACTUS. The preliminary data collection, based on a standardised approach known as NBPRA, has generated an empirical data set which is larger than any that we have so far been able to find but not yet large enough to form the basis of a reliable global parametric approach to cost estimation.

Further work is proposed to increase the data contained in the CACTUS database and start to develop heuristics to understand how key cost drivers interact to determine the relative costs of different sanitation systems in a range of contexts. The CACTUS database is intended to become a public repository for empirical sanitation cost data which will facilitate future planning.

ACKNOWLEDGEMENTS

This paper would not have been possible without the exceptional support of Anna Sevillano Evans (annasevillano98@gmail.com) in development tables and figures, and Ruthie Rosenberg (ruthie.rosenberg@saner.gy) in classifying alternative sanitation components.

DATA AVAILABILITY STATEMENT

All relevant data are included in the paper or its Supplementary Information.

REFERENCES

AACE 2019 *Cost Estimate Classification System – As Applied in Engineering, Procurement and Construction for the Process Industries (International Recommended Practice No. No. 18R-97).*

Ansar, A., Flyvbjerg, B., Budzier, A. & Lunn, D. 2014 Should we build more large dams? The actual costs of hydropower megaproject development. *Energy Policy* **69**, 43–56. https://doi.org/10.1016/j.enpol.2013.10.069.

Arnaboldi, M., Azzone, G. & Giorgino, M. 2014 *Performance Measurement and Management for Engineers*. Academic Press.

Barakchi, M., Torp, O. & Belay, A. M. 2017 Cost estimation methods for transport infrastructure: a systematic literature review. In *Procedia Engineering, Creative Construction Conference 2017, CCC 2017*, 19–22 June 2017, Primosten, Croatia. Vol. 196, pp. 270–277. https://doi.org/10.1016/j.proeng.2017.07.199.

Burr, P. & Fonseca, C. 2011 *Applying the Life-Cycle Costs Approach to Sanitation*. IRC International Water and Sanitation Centre, The Hague, Netherlands. WASH COST Briefing Note 3.

Burr, P. & Fonseca, C. 2012 *Applying the Life-Cycle Costs Approach to Water: Costs and Service Levels in Andhra Pradesh (India), Burkina Faso, Ghana and Mozambique*. IRC International Water and Sanitation Centre, La Haye.

Choudhury, S. & Saha, A. K. 2017 Prediction of operation efficiency of water treatment plant with the help of multi-criteria decision-making. *Water Conservation Science and Engineering* 1–12. https://doi.org/10.1007/s41101-017-0034-2.

Cornejo, P. K., Zhang, Q. & Mihelcic, J. R. 2013 Quantifying benefits of resource recovery from sanitation provision in a developing world setting. *Journal of Environmental Management* **131**, 7–15. https://doi.org/10.1016/j.jenvman.2013.09.043.

Crocker, J., Saywell, D., Shields, K. F., Kolsky, P. & Bartram, J. 2017 The true costs of participatory sanitation: evidence from community-led total sanitation studies in Ghana and Ethiopia. *Science of the Total Environment* **601–602**, 1075–1083. https://doi.org/10.1016/j.scitotenv.2017.05.279.

Cronin, A. A., Ohikata, M. & Kumar, M. 2014 Social and economic cost-benefit analysis of sanitation in Odisha State, India. *Journal of Water, Sanitation and Hygiene for Development* **4**, 521–531. https://doi.org/10.2166/washdev.2014.150.

Daudey, L. 2017 The cost of urban sanitation solutions: a literature review. *Journal of Water Sanitation and Hygiene for Development*. washdev2017058. https://doi.org/10.2166/washdev.2017.058.

De Roo, G. & Parsons, J. E. 2011 A methodology for calculating the levelized cost of electricity in nuclear power systems with fuel recycling. *Energy Economics* **33**, 826–839. https://doi.org/10.1016/j.eneco.2011.01.008.

De Schepper, S., Haezendonck, E. & Dooms, M. 2015 Understanding pre-contractual transaction costs for public–private partnership infrastructure projects. *International Journal of Project Management* **33**, 932–946. https://doi.org/10.1016/j.ijproman.2014.10.015.

Eggimann, S. 2016 *The Optimal Degree of Centralisation for Wastewater Infrastructures. A Model-Based Geospatial Economic Analysis (Doctoral Thesis)*. ETH Zurich. https://doi.org/10.3929/ethz-a-010811248

Eggimann, S., Truffer, B. & Maurer, M. 2015 To connect or not to connect? Modelling the optimal degree of centralisation for wastewater infrastructures. *Water Research* **84**, 218–231. https://doi.org/10.1016/j.watres.2015.07.004.

Eggimann, S., Truffer, B. & Maurer, M. 2016a The cost of hybrid waste water systems: a systematic framework for specifying minimum cost-connection rates. *Water Research* **103**, 472–484. https://doi.org/10.1016/j.watres.2016.07.062.

Eggimann, S., Truffer, B. & Maurer, M. 2016b Economies of density for on-site waste water treatment. *Water Research* **101**, 476–489. https://doi.org/10.1016/j.watres.2016.06.011.

Eggimann, S., Mutzner, L., Wani, O., Schneider, M. Y., Spuhler, D., Moy de Vitry, M., Beutler, P. & Maurer, M. 2017 The potential of knowing more: a review of data-driven urban water management. *Environmental Science & Technology* **51**, 2538–2553. https://doi.org/10.1021/acs.est.6b04267.

Eliasson, J. & Fosgerau, M. 2013 Cost overruns and demand shortfalls – deception or selection? *Transportation Research Part B: Methodological* **57**, 105–113. https://doi.org/10.1016/j.trb.2013.09.005.

Evans, B. 1995 *Institutional Incentives and the Provision of Urban Sanitation: Can the Community Increase the Chances of Success?*

Evans, B. & Mara, D. D. 2011 *Sanitation & Water Supply in Low-Income Countries*. Bookboon.

Evans, B., Voorden, C. & Peal, A. 2009 *Public Funding for Sanitation, the Many Faces of Sanitation Subsidies, Water Supply and Sanitation Collaborative Council*. Oxford Publications, London.

Evans, B., Hueso, A., Johnston, R., Norman, G., Pérez, E., Slaymaker, T. & Trémolet, S. 2017 Limited services? The role of shared sanitation in the 2030 agenda for sustainable development. *Journal of Water, Sanitation and Hygiene for Development* **7**, 349–351. https://doi.org/10.2166/washdev.2017.023.

Flyvbjerg, B. 2008 Curbing optimism bias and strategic misrepresentation in planning: reference class forecasting in practice. *European Planning Studies* **16**, 3–21. https://doi.org/10.1080/09654310701747936.

Flyvbjerg, B., Ansar, A., Budzier, A., Buhl, S., Cantarelli, C., Garbuio, M., Glenting, C., Holm, M. S., Lovallo, D., Lunn, D., Molin, E., Rønnest, A., Stewart, A. & van Wee, B. 2018 Five things you should know about cost overrun. *Transportation Research Part A: Policy and Practice* **118**, 174–190. https://doi.org/10.1016/j.tra.2018.07.013.

Fonseca, C., Dubé, A. & Verhoeven, J. 2011 *Cost-based Decision Support Tools for Water and Sanitation*. IRC International Water and Sanitation Centre, The Hague, The Netherlands.

Gatti, S. 2018 *Project Finance in Theory and Practice: Designing, Structuring, and Financing Private and Public Projects*, 3rd edn. Academic Press, S.l.

Geissmann, T. & Ponta, O. 2017 A probabilistic approach to the computation of the levelized cost of electricity. *Energy* **124**, 372–381. https://doi.org/10.1016/j.energy.2017.02.078.

Hernández-Chover, V., Bellver-Domingo, Á. & Hernández-Sancho, F. 2018 Efficiency of wastewater treatment facilities: the influence of scale economies. *Journal of Environmental Management* **228**, 77–84. https://doi.org/10.1016/j.jenvman.2018.09.014.

Hughes, W., Champion, R. & Murdoch, J. 2015 *Construction Contracts: Law and Management*, 5th edn. Routledge, Abingdon.

Hutton, G. 2013 Global costs and benefits of reaching universal coverage of sanitation and drinking-water supply. *Journal of Water and Health* **11**, 1–12. https://doi.org/10.2166/wh.2012.105.

Hutton, G. & Bartram, J. 2008 *Regional and Global Costs of Attaining the Water Supply and Sanitation Target (Target 10) of the Millennium Development Goals*. World Health Organization, Geneva.

Hutton, G. & Haller, L. 2004 *Evaluation of the Costs and Benefits of Water and Sanitation Improvements at the Global Level*. World Health Organization, Geneva.

Hutton, G., Rodriguez, U.-P., Winara, A., Anh, N. V., Phyrum, K., Chuan, L., Blackett, I. & Weitz, A. 2014 Economic efficiency of sanitation interventions in Southeast Asia. *Journal of Water, Sanitation and Hygiene for Development* **4**, 23–36. https://doi.org/10.2166/washdev.2013.158.

Ika, L. A. & Donnelly, J. 2017 Success conditions for international development capacity building projects. *International Journal of Project Management* **35**, 44–63. https://doi.org/10.1016/j.ijproman.2016.10.005.

Ika, L. A., Diallo, A. & Thuillier, D. 2012 Critical success factors for World Bank projects: an empirical investigation. *International Journal of Project Management* **30**, 105–116. https://doi.org/10.1016/j.ijproman.2011.03.005.

International Water Association 2014 *Human Resource Capacity Gaps Study [WWW Document]*. International Water Association. Available from: http://www.iwa-network.org/projects/human-resource-capacity-gaps-study/ (accessed 11 July 2018).

Kalbermatten, J. M., Julius, D. S., Gunnerson, C. G. & Mara, D. D. 1982 *Appropriate Sanitation Alternatives: A Planning and Design Manual, World Bank Studies in Water Supply and Sanitation*. Published for the World Bank [by] the Johns Hopkins University Press, Baltimore.

Ketema, A. A. & Langergraber, G. 2016 Statistical validation of the CLARA simplified planning tool. *Water Science and Technology: Water Supply* **16**, 193–201. https://doi.org/10.2166/ws.2015.125.

Ketema, A. A., Lechner, M., Tilahun, S. A. & Langergraber, G. 2015 Development of cost functions for water supply and sanitation technologies: case study of Bahir Dar and Arba Minch, Ethiopia. *Journal of Water Sanitation and Hygiene for Development* **5**, 502–511. https://doi.org/10.2166/washdev.2015.067.

Kim, G.-H., An, S.-H. & Kang, K.-I. 2004 Comparison of construction cost estimating models based on regression analysis, neural networks, and case-based reasoning. *Building and Environment* **39**, 1235–1242. https://doi.org/10.1016/j.buildenv.2004.02.013.

Lai, C. S. & McCulloch, M. D. 2017 Levelized cost of electricity for solar photovoltaic and electrical energy storage. *Applied Energy* **190**, 191–203. https://doi.org/10.1016/j.apenergy.2016.12.153.

Lakner, C., Mahler, D. G., Nguyen, M. C., Azevedo, J. P., Chen, S., Jolliffe, D., Prydz, E. B. & Sangraula, P. 2018 *Consumer Price Indices Used in Global Poverty Measurement*. Global Poverty Monitoring Technical Note.

Locatelli, G. 2018 *Why are Megaprojects, Including Nuclear Power Plants, Delivered Overbudget and Late? Reasons and Remedies*. Center for Advanced Nuclear Energy Systems (CANES), Massachusetts Institute of Technology.

Locatelli, G., Mariani, G., Sainati, T. & Greco, M. 2017 Corruption in public projects and megaprojects: there is an elephant in the room! *International Journal of Project Management* **35**, 252–268. https://doi.org/10.1016/j.ijproman.2016.09.010.

Loetscher, T. & Keller, J. 2018 *Estimating the Financial Costs of Sanitation Systems*.

Loutatidou, S., Chalermthai, B., Marpu, P. R. & Arafat, H. A. 2014 Capital cost estimation of RO plants: GCC countries versus southern Europe. *Desalination* **347**, 103–111. https://doi.org/10.1016/j.desal.2014.05.033.

Love, P. E. D., Ahiaga-Dagbui, D. D. & Irani, Z. 2016 Cost overruns in transportation infrastructure projects: sowing the seeds for a probabilistic theory of causation. *Transportation Research Part A: Policy and Practice* **92**, 184–194. https://doi.org/10.1016/j.tra.2016.08.007.

Love, P. E. D., Sing, M. C. P., Ika, L. A. & Newton, S. 2019 The cost performance of transportation projects: the fallacy of the Planning Fallacy account. *Transportation Research Part A: Policy and Practice* **122**, 1–20. https://doi.org/10.1016/j.tra.2019.02.004.

Luh, J., Royster, S., Sebastian, D., Ojomo, E. & Bartram, J. 2017 Expert assessment of the resilience of drinking water and sanitation systems to climate-related hazards. *Science of The Total Environment* **592**, 334–344. https://doi.org/10.1016/j.scitotenv.2017.03.084.

Lutterbeck, C. A., Kist, L. T., Lopez, D. R., Zerwes, F. V. & Machado, Ê. L. 2017 Life cycle assessment of integrated wastewater treatment systems with constructed wetlands in rural areas. *Journal of Cleaner Production* **148**, 527–536. https://doi.org/10.1016/j.jclepro.2017.02.024.

Mara, D. 1996 *Low-Cost Sewerage*. John Wiley.

Mara, D. & Evans, B. 2018 The sanitation and hygiene targets of the sustainable development goals: scope and challenges. *Journal of Water, Sanitation and Hygiene for Development* **8**, 1–16. https://doi.org/10.2166/washdev.2017.048.

Mara, D. D. & Guimarães, A. S. P. 1999 Simplified sewerage: potential applicability in industrialized countries. *Urban Water* **1**, 257–259. https://doi.org/10.1016/S1462-0758(99)00015-1.

McIntyre, P., Casella, D., Fonseca, C. & Burr, P. 2014 *Priceless!: Uncovering the Real Costs of Water and Sanitation*. IRC.

Merrow, E. W. 2011 *Industrial Megaprojects: Concepts, Strategies, and Practices for Success*, 1 edn. John Wiley & Sons, Hoboke, NJ.

Mislick, G. K. & Nussbaum, D. A. 2015 *Cost Estimation: Methods and Tools*. John Wiley & Sons.

Mitlin, D. 2015 Will urban sanitation 'leave no one behind'? *Environment and Urbanization* **27**, 365–370. https://doi.org/10.1177/0956247815604527.

Müllegger, E., Langergraber, G. & Lechner, M. 2014 *Issue 19 -The CLARA Project*. Sustainable Sanitation Practice.

OECD 2007 Cost estimating guidelines for generation iv nuclear energy systems. In *The Economic Modeling Working Group Of the Generation IV International Forum*.

Peal, A., Evans, B., Ahilan, S., Ban, R., Blackett, I., Hawkins, P., Schoebitz, L., Scott, R., Sleigh, A., Strande, L. & Veses Roda, O. 2020 *Estimating Safely Managed Sanitation in Urban Areas; Lessons Learned From A Global Implementation of Excreta-Flow Diagrams*. https://doi.org/10.3389/fenvs.2020.00001.

Perard, E. 2018 Economic and financial aspects of the sanitation challenge: a practitioner approach. *Utilities Policy* **52**, 22–26. https://doi.org/10.1016/j.jup.2018.03.007.

Rao, D. P. 2001 *Integration of CPI and PPP: Methodological Issues, Feasibility and Recommendations*. Joint World Bank–OECD Seminar on PPPs.

Roman, W., Weil, J. F. & Schipper, K. 2012 *Financial Accounting: An Introduction to Concepts, Methods and Uses*, 14th edn. Cengage Learning.

Ross, I., Scott, R., Mujica, A., White, Z. & Smith, M. 2016 *Fecal Sludge Management: Diagnostics for Service Delivery in Urban Areas – Tools and Guidelines (Water and Sanitation Program Technical Paper), Water and Sanitation Program (WSP)*. World Bank Group, Washington, DC.

Ruiters, C. & Matji, M. P. 2017 Funding and financing for wastewater and sanitation infrastructure in South Africa: pricing, tariffs and operational efficiency. In *Infrastructure Investments: Politics, Barriers and Economic Consequences*. pp. 183–215.

Sinnatamby, G. S., Mara, D. D. & McGarry, M. G. 1986 Shallow systems offer hope to slums. *World Water* **9**, 39–41.

Stewart, R. D., Wyskida, R. M. & Johannes, J. D. 1995 *Cost Estimator's Reference Manual*. John Wiley & Sons, New York.

Tilley, E., Ulrich, L., Lüthi, C., Reymond, P., Schertenleib, R. & Zurbrügg, C. 2014 *Compendium of Sanitation Systems and Technologies*, 2nd edn. Swiss Federal Institute of Aquatic Science and Technology (Eawag), Dübendorf.

UNICEF, WHO 2019 *Progress on Household Drinking Water, Sanitation and Hygiene 2000–2017: Special Focus on Inequalities*. World Health Organization.

Whittington, D., Jeuland, M., Barker, K. & Yuen, Y. 2012 Setting priorities, targeting subsidies among water, sanitation, and preventive health interventions in developing countries. *World Development* **40**, 1546–1568. https://doi.org/10.1016/j.worlddev.2012.03.004.

Williamson, O. E. 1979 Transaction-cost economics: the governance of contractual relations. *The Journal of Law and Economics* **22**, 233–261. https://doi.org/10.1086/466942.

Williamson, O. E. 1981 The economics of organization: the transaction cost approach. *American Journal of Sociology* **87**, 548–577. https://doi.org/10.1086/227496.

World Bank 2020 *Data Bank Webisete*.

World Health Organization 2012 *Global Costs and Benefits of Drinking-Water Supply and Sanitation Interventions to Reach the MDG Target and Universal Coverage (No. WHO/HSE/WSH/12.01)*. World Health Organization.

Zurbrügg, C., Caniato, M. & Vaccari, M. 2014 How assessment methods can support solid waste management in developing countries – a critical review. *Sustainability* **6**, 545–570. https://doi.org/10.3390/su6020545.

First received 3 May 2020; accepted in revised form 2 October 2020. Available online 27 October 2020

Research Paper

Life-cycle costs approach for private piped water service delivery: a study in rural Viet Nam

Melita Grant, Tim Foster, Dao Van Dinh, Juliet Willetts and Georgia Davis

ABSTRACT

Sustaining rural water services in Viet Nam requires an improved understanding of the costs and relative opportunities, especially given the government's support for private sector involvement in expanding water schemes. In particular, the life-cycle costs associated with the delivery of safe and sustainable water services in rural Viet Nam, as indeed elsewhere, are not well known, potentially compromising their long-term sustainability. To address this gap, this study assessed the cost structures of 14 water schemes in Viet Nam managed by private enterprises. Results showed that both capital and operational expenditures varied widely across the schemes assessed, reflective of the diversity of the age and characteristics of the schemes studied. Twelve of the 14 schemes generated a cash profit in the most recent calendar year; however, when taking into account depreciation, as well as historical subsidies and connection fee payments, only four of the schemes were profitable based on a 20-year design life assumption. The study complements previous research demonstrating barriers to achieving universal access when relying on user-pays systems. The results provide a useful reference point to inform business planning for enterprises, as well as policy and support mechanisms important for securing sustainable rural water supply services.

Key words | costs, enterprises, finance, life-cycle, private, water

Melita Grant (corresponding author)
Tim Foster
Juliet Willetts
Institute for Sustainable Futures, University of Technology Sydney,
235 Jones Street, Ultimo, NSW 2007,
Australia
E-mail: *melita.grant@uts.edu.au*

Dao Van Dinh
Institute for Economics and Water Resources Management (IWEM)
131 Chua Boc, Dong Da, Ha Noi,
Vietnam

Georgia Davis
East Meets West Foundation (EMWF),
No. 2, Lane 10, Dang Thai Mai Street, Quang An Ward, Tay Ho District, Ha Noi,
Vietnam

HIGHLIGHTS

- This study assessed cost structures for 14 privately operated water supply schemes in rural Viet Nam.
- Capital expenditure varied significantly, from USD 55 to USD 522 per connection and operational expenditures were driven by labour and electricity costs.
- Reported expenditure on asset renewal was low, which could compromise long-term service sustainability.
- Only four of the 14 schemes appeared financially sustainable based on a 20-year design life assumption, requiring attention to life-cycle costs informed financial management.

INTRODUCTION

The rural water sector in low- and middle-income countries (LMICs) has been characterised by a move away from an infrastructure-focused model to a service delivery model in recent years (Moriarty *et al.* 2013), though the extent to which this shift is taking place in reality remains unclear. This shift has been spurred by the growing recognition of non-functional and unsustainable systems, hindering the realisation of the health and welfare benefits that would normally be expected to flow from improvements in water supply infrastructure (Burr & Fonseca 2013). A service delivery model requires the

doi: 10.2166/washdev.2020.037

full costs of WASH services to be better understood and managed, so that systems can be maintained, potentially expanded and sustained in the long term (World Bank 2017).

The required costs and investment to expand and sustain services is a critical issue facing the sector. The higher bar set by the Sustainable Development Goals (SDGs) for targets 6.1 and 6.2 requires countries to aim for 'safely managed' services rather than provision of access to 'improved' infrastructure alone. 'Safely managed' in the water services context means that water supplies must be located on premises, available when needed and free of faecal and priority chemical contamination (UN 2017). Hutton & Varughese (2016) found that capital investments required to achieve SDG targets 6.1 and 6.2 amount to about three times current investment levels, and that financial and institutional strengthening will be critical for ensuring any increase in capital investment leads to sustainable services. This research contributes to the broader call for strengthening the enabling environment through an improved understanding of the real costs associated with delivering sustainable water supply services in rural areas, so that these costs can be met through a range of sources. The WASH enabling environment has been defined as a 'set of interrelated sector functions that enable governments and public and private partners to engage in a sustained and effective WASH service delivery development process' (UNICEF 2016, p. ix).

There is growing support for the life-cycle cost approach (LCCA) as a tool for driving sustainability improvements in the rural water supply sector, including in Viet Nam. In a water service context, life-cycle costs refer to the full set of costs associated with delivering an adequate water service indefinitely (Fonseca *et al.* 2010). For our study, we included capital expenditure, operating and minor maintenance expenditure, capital maintenance expenditure and cost of capital. We did not include expenditure on direct support and expenditure on indirect support (see Supplementary Material, Table S1). The application of LCCA to WASH in countries in the Global South is nascent, but has been spurred by the WASHCost initiative, which aimed to make LCCA methodologies more accessible to the rural WASH sector (Fonseca *et al.* 2011). Few peer-reviewed studies have analysed revenue and expenditure (costs) of small-scale rural water services in LMICs. Rather, LCCA has primarily been used for urban systems in well-resourced countries (Jones *et al.* 2012). Prior to the present study, LCCA had yet to be applied to rural water services in Viet Nam and is not known to have been applied to privately operated rural water services in LMICs more generally. This is despite growing acknowledgement among policy-makers of the need to better understand the true costs associated with providing rural water services. Illustrative of this is Viet Nam's National Action Plan, which articulates a desire for the development of regulations on cost norms (referred to as technical-economic standards in Viet Nam) for clean water supply in rural areas in pursuit of SDG target 6.1 (Government of Viet Nam 2017). Without rigorous life-cycle cost analysis of rural water supplies, sustainability improvements in the rural water sector will not be realised.

In the Global South, implementing an LCCA for water scheme options is difficult because there is limited available data and water system design is dependent on the local context (Jones *et al.* 2012). One study assessed water provision costs in 43 villages in Andhra Pradesh, India, finding that unit costs were substantially higher than the prescribed cost norms, and that norms were unrealistic (Reddy *et al.* 2012). Reddy *et al.* (2012) also found that the allocations to rural drinking water were low at the design and implementation stage, which resulted in more needing to be spent later on due to *ad hoc* interventions. Another study on primary schools in Kenya estimated current expenditures on WASH to be only 60% of requirements to meet minimum standards and recurrent costs (Alexander *et al.* 2016). A study of rural water supplies in Kenya analysed water committee records and found expenditure levels varied greatly across schemes, with both operational model and seasonality influencing the magnitude and timing of operation and maintenance costs (Foster & Hope 2017). These studies show LCCA to be useful to identify financing gaps and economic sustainability issues which, if unaddressed, can lead to non-functional or under-performing water schemes.

Background: privately operated piped water services in Viet Nam

Rural Viet Nam has seen rapid growth in piped water service coverage in recent years due to a supportive policy environment. From 2000 to 2016, around 16,200 piped water schemes were constructed in rural areas (Mansour

2016), increasing piped water access from one million people in 2000 (1.7% of the rural population) to 13.7 million people (22.1% of the rural population) in 2015 (UNICEF/WHO 2020). This trajectory is likely to continue, with Vietnam's Country Strategy on Rural Water Supply and Sanitation to 2020 setting the target that by 2020: 'all rural residents use clean water meeting national standards, and access at least 60 litres/person/day' (CERWASS 2000). The elevated SDG ambition of safely managed water services, with its requirement of being 'on premises' and free from contamination, provides further impetus to accelerate the expansion of piped water systems.

Government and development actors in Viet Nam have increasingly promoted private sector management models through a range of policy levers and funding support. According to the Viet Nam Government's Ministry for Agriculture and Development (MARD), private enterprises manage 13% of rural water supply systems in Viet Nam, with the community managing 70% and the Provincial Centre for Rural Water Supply and Sanitation (pCERWASS) and state boards managing 17% of schemes (Directorate of Water Resources 2018). In the Vietnamese context, private water enterprises are entities that have invested private funds in a water system and own and/or operate the system under a formal or informal agreement with a Provincial People's Committee (PPC) or a Commune People's Committee (CPC). Policy support for private sector involvement in the delivery of rural water services in Viet Nam dates back almost two decades. The National Rural Clean Water Supply and Sanitation Strategy 2020 (developed in 2000 and updated in 2011) stated that the Viet Nam Government would encourage the private sector to invest in and to construct rural water facilities, especially piped water supply systems. This encouragement extends to taking over from government and community-based schemes, as well as the establishment of new schemes in areas that are currently unserved.

Despite the policy incentives, there is limited understanding of the financial viability of private water enterprises (Willetts *et al.* 2017). This is an important issue to clarify in light of government data that suggests approximately 14.4% of rural piped systems in Viet Nam are not working, and a further 16.8% are performing poorly (Directorate of Water Resources 2018). As noted by Moriarty *et al.* (2013), the first step to ensuring the financial sustainability of services lies in identifying the full life-cycle costs of service provision and the revenue sources that can meet those costs.

This study collected empirical cost data and identified how those costs are met and allocated between different stakeholder groups. This includes, for example, how much of the total cost is passed on to customers, to what extent services are subsidised, by whom and for whom. The broader purpose of the study was to provide the Government of Viet Nam (at Central, Provincial and Commune level) and its development partners a better understanding of the costs and revenues of small-scale piped water systems over time. The findings can support government agencies to develop evidence-based policies, regulations and incentive structures, aid private enterprises in business planning and structuring connection fees and tariffs, and inform efforts of development partners to design and implement targeted support mechanisms.

METHODS

This investigation adapted the LCCA process developed by WASHCost (Fonseca *et al.* 2011) to quantify the costs associated with delivering piped water services for 14 schemes in rural Viet Nam (see Supplementary Material, Figure S1). The study also identified the factors that drive and influence those costs and assessed the ability of scheme operators to cover those costs. The LCCA included the aggregate financial costs for service providers across capital expenditure, operating and minor maintenance expenditure, capital maintenance expenditure and cost of capital. As the research scope was limited to the costs incurred by the service providers themselves, no data were collected on direct support (capacity of local government to regulate) or indirect support (planning and policy development) expenditure, as would ideally be included in the WASHCost methodology (see Supplementary Table S1 for further explanation).

Financial data were collected from 14 privately owned and operated piped water supply schemes between December 2016 and March 2018. The data collection tool was piloted in December 2016 with two schemes and then refined. The sites were selected through a purposive sampling process, with four located in the northern part of the country and ten in the South. This geographical representation was considered important given variability in technical, institutional and hydrogeological characteristics. For example, in the North of Viet Nam, private enterprises are often larger than those in the South/Mekong region and draw from surface water sources (as opposed to groundwater) to a greater extent. The original research design included equal numbers of schemes in the North and the South; however, due to unforeseen human resources constraints, only four schemes were able to be studied in the North of Viet Nam.

Data collection instruments were designed to capture a comprehensive set of costs associated with construction, operation and maintenance of small-scale piped water services. The structure and content of the questionnaire were informed by the WASHCost LCCA framework. Additional questions on technical and operational characteristics (e.g., size, service level and water source) were also included.

Data on the various costs, revenues and system characteristics were provided by the individual responsible for the overall management of each scheme. The interview process was conducted in Vietnamese and typically lasted half a day or more. Financial information was drawn from various sources, including accounting records, invoices and respondent recall, which is a limitation of the study (see Supplementary Material for additional explanation of limitations which include loss of documentation, selection of schemes, age of schemes impacting on availability of cost documentation, issues related to per connection calculations, due to large industrial or institutional customers not being able to be accounted for, wide range of schemes in terms of size and type, and potential unwillingness of scheme owners to share sensitive financial information). Follow up was undertaken where data were ambiguous, anomalous or missing. Ethics approval for the study was obtained from the University of Technology Sydney.

For each scheme, individual cost items were aggregated and assessed for each category as defined by the WASHCost methodology. Operational expenditure (OpEx), cost of capital (CoC) and capital maintenance expenditure (CapManEx) were annualised for consistency, and all costs were converted to US dollars in 2016 by applying deflator factors and a period average exchange rate of USD 1 equal to VND 21,935 (World Bank 2018a, 2018b). Costs were overlayed with reported revenue to assess the profitability of schemes. To compare the WASHCost LCCA methodology with Viet Nam government-endorsed accounting practices for water service providers, annual depreciation was calculated based on a straight-line method applied to capital expenditure. To identify major cost drivers, OpEx was disaggregated into five key categories: labour, electricity, chemicals, taxes and 'other'. Average costs were also compared across different factors and dimensions, including water source, season, scheme size, age and service level.

RESULTS

Scheme characteristics

While all schemes included in the study were rural piped water schemes, they had diverse characteristics. The 14 schemes varied in age, size and production capacity (Table 1 and Supplementary Material, Tables S3–S7 for more detail). The age of the schemes ranged from 4 to 21 years and on average supplied water for 20 h/day. The average tariff for each connected property (across all property types) ranged from USD 0.26 to USD 0.38, and average connection fee was US$53. Half of the schemes drew on surface

Table 1 | Summary characteristics of the schemes included in the research ($n = 14$)

Attribute	Mean	Standard deviation	Min	Max
System age (years)	14.2	6.3	4	21
Hours of service per day	19.5	5.4	7	24
Number of connections	1,633	986	720	3,555
Production capacity (m^3/day)	1,144	783	120	2,400
Water supplied per connection (l/day)[a]	311	136	125	556
Tariff (USD/m^3)	0.32	0.04	0.26	0.38
Connection fee charged (USD)	53	49	0	153

[a]Note that the number of connections includes business and community institutions (hotels, schools and water consuming businesses), and so, this does not reflect domestic use per connection.

water, and the other half drew on groundwater. Based on WASHCost definitions of water service levels, the schemes typically performed to a high standard against quantity and accessibility, to an intermediate to high standard in terms of reliability, while the extent to which schemes provided water of acceptable quality was unclear. It is important to note that these service standards were self-reported by scheme operators and were not independently verified. Within our sample, there was little variability in service standards as reported by the operators, with 13 of the 14 schemes reported to work most of the time with an average of 19.5 h of service per day (see Supplementary Material, Table S2).

Capital expenditure

Capital expenditure (CapEx) reported by scheme operators averaged USD 324 per connection and varied between USD 55 and USD 522 per connection (Table 2, and Table S3 in Supplementary Material). Around one-quarter of capital investment related to system expansions subsequent to initial construction. For those schemes reporting the source of financing, 80% of capital investment was financed by the scheme owners either directly or via loans taken out from a bank/lender, 9% by NGOs, 7% by government and 4% by households. However, extrapolation of water connection fees suggests up to 20% of capital investment may have

Table 2 | Mean expenditure reported by schemes across key cost categories and scheme characteristics

		Capex (cumulative)		Opex (annual)			CapManEx (annual)			CoC (annual)		
Characteristic	***n***	**Total**	**Per connection**	**Total**	**Per connection**	**Per m^3**	**Total**	**Per connection**	**Per m^3**	**Total**	**Per connection**	**Per m^3**
All schemes	14	524,208	324	33,038	23.9	0.22	2,652	2.60	0.02	1,462	0.98	0.01
Source type												
Groundwater	7	386,719	274	28,871	18.9	0.22	2,406	1.75	0.02	2,019	1.02	0.01
Surface water	7	661,698	374	37,206	28.8	0.23	2,898	3.44	0.02	905	0.95	0.01
Scheme size												
<1,500 connections	9	317,327	337	26,866	27.7	0.26	3,202	3.66	0.03	907	0.94	0.01
≥1,500 connections	5	896,594	302	44,148	16.9	0.16	1,662	0.68	0.01	2,462	1.07	0.01
Treatment												
Yes	11	519,170	316	33,985	25.4	0.22	2,897	3.04	0.02	1,861	1.26	0.01
No	3	542,680	354	29,569	18.1	0.23	1,755	0.96	0.02	0	0.00	0.00
Service continuity[a]												
<24 h supply	7	765,609	426	30,778	20.6	0.22	3,226	3.50	0.03	423	0.49	0.00
24 h supply	6	251,458	217	33,629	27.5	0.24	2,207	1.86	0.02	2,918	1.74	0.02
Scheme age												
<10 years	5	821,032	380	35,206	16.3	0.19	1,310	0.86	0.01	2,827	1.43	0.01
≥10 years	9	359,306	294	31,834	28.0	0.24	3,398	3.56	0.03	704	0.74	0.01
Region												
North	4	936,455	471	36,632	23.7	0.20	3,576	4.61	0.03	285	0.40	0.00
South	10	359,310	266	31,601	23.9	0.23	2,283	1.79	0.02	1,933	1.23	0.01
Tariff (USD/m^3)												
<0.31	7	722,704	355	34,516	20.6	0.20	3,013	3.13	0.02	163	0.23	0.00
≥0.31	7	325,712	294	31,561	27.1	0.24	2,291	2.06	0.02	2,762	1.75	0.02

[a]Information on service continuity could not be obtained from one scheme.

been covered by households. External subsidies constituted a larger proportion of system upgrades and expansions compared with initial construction cost (27% vs. 10%). Schemes that were younger than 10 years tended to report higher levels of CapEx per connection (USD 380 vs. 294), as did schemes drawing on surface water as opposed to groundwater (USD 374 vs. 274), and schemes without a water treatment plant (USD 354 v. 316). (Based on the information provided by water scheme owners, three of the schemes had no centralised treatment.) Larger schemes (≥1,500 connections) reported slightly lower CapEx per connection than smaller schemes (USD 302 vs. 337).

Recurrent expenditure

Operational expenditure (OpEx) reported by scheme operators averaged USD 23.9 (range USD 11–48.4) per connection per year and USD 0.22 m^{-3} (range USD 0.09–0.34 m^{-3}) (Table 2). Only one scheme reported unit OpEx that exceeded their volumetric tariff, indicating that OpEx was likely covered by tariffs in most of the schemes involved in this study.

Labour and electricity constituted the major operational cost categories, amounting to 43 and 25% of the average costs, respectively, followed by taxes (10%) and chemicals (9%) (Figure 1). Taxes reported by scheme operators typically included a 5% tax for environmental protection, though some reported the tax to be between 1 and 6%, calculated based on the amount of water abstracted from the system. The water resource upon which a scheme relied had an influence on OpEx: schemes lifting groundwater spent almost 41% more on electricity (per cubic meter of water supplied) than those drawing on surface water, but their chemical costs were 79% lower. Scheme size – particularly when measured by volume of water supplied – exhibited a negative association with unit operational costs ($p < 0.001$), indicative of economies of scale. OpEx incurred by schemes in the South appeared to be heavily influenced by monthly rainfall, suggestive of rainwater harvesting practices diminishing demand for piped water (Figure 2).

Scheme operators reported additional costs, albeit relatively minor, relating to capital maintenance (CapManEx) and cost of capital (interest repayments) (see Supplementary Material, Tables S5 and S6). For the CapManEx attributed to the replacement of specific infrastructure components, 53% related to pipes and meters, 19% to pumps and 16% to treatment plant. In aggregate terms, the annualised CapManEx amounted to 8% of operational expenditure, perhaps raising questions about whether recall bias may have resulted in an underestimate. Most schemes reported historical CapManEx, though the average annualised amount was relatively small (USD 2.60 per connection per year; 0.02 m^{-3}). Older schemes reported higher levels of CapManEx, with systems more than 10 years old recording CapManEx four times that of schemes under 10 years. Six schemes reported paying interest on loans; when averaged across all schemes, these payments amounted to USD 0.98 per connection per year (USD 0.01 m^{-3}).

Profitability

Twelve of the 14 schemes generated a cash profit in the most recent calendar year, though three of the profitable schemes

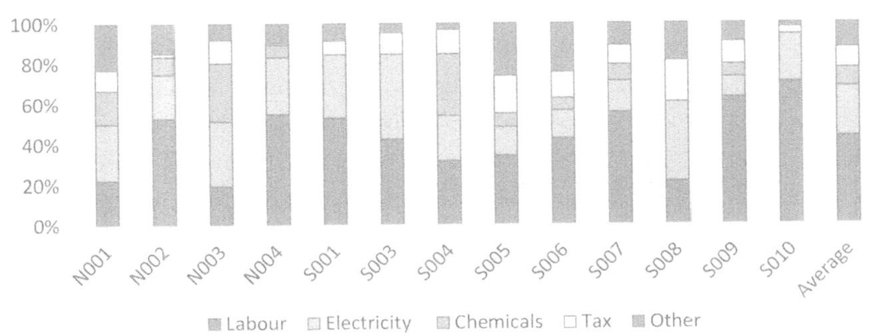

Figure 1 | Disaggregation of operational expenditure by major cost categories. *Note*: Category 'Other' includes materials, water quality monitoring, rental costs, fuel and other administrative expenses. Data did not allow for disaggregation of costs for scheme S002. N in this table denotes North, and S denotes South.

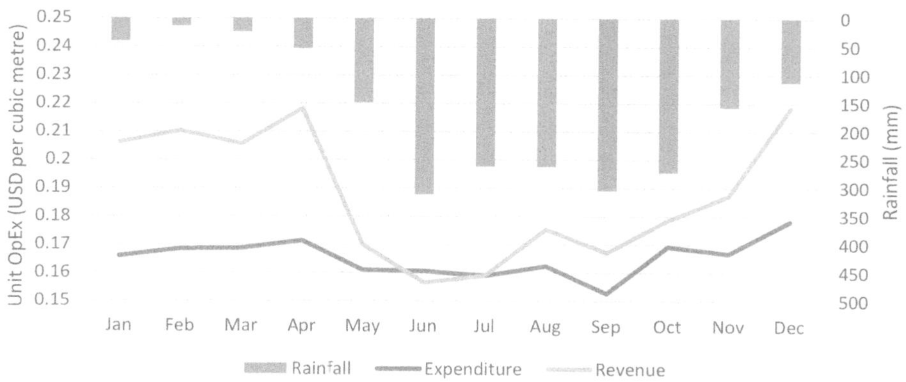

Figure 2 | Operational expenditure and rainfall by month in southern Vietnam (seven schemes in Bên Tre and Long An Province) (*Note*: This is a subset of the data that was able to be disaggregated by month (*n* = 7). Not all scheme owners were able to complete, or did not wish to share, the month by month break down of expenses and revenue).

generated a surplus of less than USD 1,000 (see Supplementary Material, Table S7). Only one scheme reported making a substantial loss, and this scheme also happened to have the lowest tariff. When taking into account depreciation, historical subsidies and connection fee payments, four of the schemes were profitable based on a 20-year design life assumption.

DISCUSSION

This study is the first to quantify a broad suite of costs associated with privately owned and managed water service delivery in rural Viet Nam. It is also the first to document service delivery costs incurred by privately owned and operated schemes in LMICs more broadly. In doing so, the analysis generated a range of new insights relating to cost variation, cost drivers, capital maintenance and depreciation, tariff determination and connection fees, evidence gaps and the financial sustainability of small-scale piped schemes.

The results demonstrate how unit life-cycle costs can vary dramatically. This variation makes it challenging to define standardised cost norms (technical-economic standards) in the Vietnamese context. Rather, if cost norms are developed, as was planned for some jurisdictions in Viet Nam, they should take into account a range of key factors and be flexible to local circumstances. Key factors which influenced costs included scheme size, water resources (source, quality and reliability), climate and service level. These multiple influences mean that generalised cost norms or benchmarks should be interpreted and employed carefully and in light of contextual factors. Moreover, formulation of such norms should be premised on a range or distribution of values that are appropriate to rural small–medium size water schemes, rather than being based on urban water utilities norms or single numerical values or averages.

Some of the drivers of cost variability in this study align with previous research, while others are novel observations. The relationship between cash flows and rainfall is similar to that documented in rural Kenya (Foster & Hope 2017), while Burr & Fonseca (2013) also noted that a drop-off in water use can impact LCCA. However, in contrast to Burr and Fonseca, this study found only modest economies of scale when it came to capital investment. Rather, it was unit operational expenditure that reduced most substantially with scheme size. Variation of costs based on context, while unsurprising, was an important finding of this study with significant policy implications for Viet Nam and other countries wishing to set cost norms to inform connection fees and tariffs.

Reported CapManEx (expenditure on asset renewal) was, on average, surprisingly low. While costing of rural piped services elsewhere have found annual CapManEx to be of the same order of magnitude as annual OpEx (WASH-Cost 2012), our results reveal annualised CapManEx to be an order of magnitude lower than annual OpEx. On the one hand, this could be a consequence of adequate levels of spending for operational expenditure (i.e., regular maintenance that prevents the need for asset replacement); on the other, it may reflect infrastructure being pushed beyond its limit in a bid to defer its renewal. The reported low levels of CapManEx could also have been a result of respondent recall issues, notwithstanding the systematic and rigorous data collection processes employed by the study. The difficulty of obtaining reliable cost data is a known challenge of conducting research into water service costs in rural areas of LMICs (Burr & Fonseca 2013), so some costs such as CapManEx are likely to have been underreported. Record keeping was at times poor, making it difficult to verify data provided. This applies particularly to historical costs that were incurred more than 12 months prior to the time of data collection, meaning the challenge of collecting robust information was most acute for CapManEx and CapEx associated with older schemes. Inadequate financial record-keeping processes by service providers were compounded by the fact that older schemes were more likely to have changed hands (either in ownership or managerially) since the initial construction. It is, therefore, likely that estimates for both of these cost categories suffered from measurement error. Implications of low CapManEx, whether it be as a result of under-reporting or under-investment, point to the need for a better understanding of how assets are being maintained or upgraded such that they can provide long-term services for the community. Our study found that insufficient documentation contributed to the inability for enterprises and authorities to calculate the life-cycle costs accurately.

Beyond the reported CapManEx collected through this study, depreciation was also calculated as a way to consider whether reported CapManEx costs were likely to be sufficient to maintain assets. Depreciation is a critical consideration, not least because it is promoted by relevant policy instruments in Viet Nam. For example, Circular 45 (Guiding regulation on management, use and depreciation of fixed assets) sets out the way depreciation is to be calculated for fixed assets, and includes an annex which identifies a minimum and maximum design life over which a range of fixed assets should be depreciated (Government of Viet Nam 2013). This study found that a simple straight-line depreciation of capital investment that assumed a design life of 20 years resulted in an average of USD 16.20 per connection per year, which was six times greater than the CapManEx estimate. Again, this may point to under-investment, under-reporting of CapManEx or a combination of the two. Whatever the case, the findings point to a broader maintenance and financing challenge that water service providers will need to overcome in Viet Nam and beyond.

Most of the schemes appeared to be cash flow positive in the previous calendar year. However, given the low level of reported CapManEx, this may not have been reflective of the life-cycle costs required to ensure ongoing sustainability of the scheme. While some of the schemes had been in place for more than a decade, their continued operation may have only been possible due to subsidies provided by government or development partners in previous years. When depreciating capital investment using a straight-line method and factoring in historical subsidies and connection fee payments, only four schemes appeared to be profitable under a 20-year design life assumption. The study found that older schemes had much higher CapManEx which demonstrated that these costs can be expected to increase over the life of a system (unless proactive maintenance is undertaken). The results suggest that subsidies will be needed to ensure ongoing commercial viability in the face of depreciating assets and related capital maintenance costs, and to reach people who are poor and/or marginalised. Additionally, the viability of schemes could be improved by adjusting tariffs, taking into account equity and obligations under the Human Rights for Water and Sanitation, different uses of water (e.g., commercial and domestic use), to ensure affordable and sustainable services for all.

Tariff levels appeared to be adequate to cover operational expenditure, but not likely the full life-cycle costs. Tariffs for piped water supply in Viet Nam are capped by provincial authorities, and in the 14 schemes included in this study, the tariff ranged from USD 0.26 to USD 0.38 m^{-3}. As the profitability analysis suggests, these were generally sufficient to cover operation and maintenance costs, and in some cases (33% of schemes studied) capital investment too. In Viet Nam, there are various policy instruments in place to encourage pricing to be set with reference to real costs (such as Circular 75, 2012), but these instruments have been found not to be applicable to rural water supply and sanitation contexts, given their focus on larger urban utilities. The prescribed tariff setting methodology in the water supply sector globally considers a range of key principles including that the user pays; cost-recovery; the principle of the human right to water based on UN Resolution UN 64/292; that the price is affordable and if not, then subsidies are offered; water conservation and natural resources management and economic efficiency (Vucijak *et al.* 2015). In this study, there was little evidence that the tariff rate charged to customers had been set with consideration of the key principles outlined above, and in some provinces, tariffs were not tailored to each scheme based on its unique costs, but the maximum tariff allowed was charged. Questions remain as to whether or not CapManEx is adequately resourced; however, if such costs are factored in to the life-cycle costs, then tariff levels may not be sufficient to cover all costs associated with a small-scale piped water scheme. To further clarify this area, systems for tracking financial flows would need to be put in place to ascertain the full range of costs over the longer term, with consideration of how to ensure tariffs remain affordable for all.

Connection fees are an important source of revenue for private enterprises to meet the costs faced in early phases of establishment of their business. Connection fees in this study were found to vary from USD 153 to no fee. Some schemes offered free or low-cost connections if households were located close to an existing pipeline or were deemed to be poor. Research on equity dimensions of small-scale private water schemes in Viet Nam revealed that connection fees were a key barrier for poor households to connect to piped water services in rural contexts (Grant *et al.* 2016a, 2016b; Carrard *et al.* 2019). Further consideration of how connection fees contribute to overall revenue sources and commercial viability for enterprises will be important for understanding any related requirement for government subsidies to avoid exclusion of the disadvantaged.

Life-cycle costs vary across country and local contexts, and relate strongly to the relevant water supply service level. Both the OpEx and CapEx reported in this study were substantially higher than those documented by the WASHCost initiative in Burkina Faso, Ghana, Mozambique and India (Burr & Fonseca 2013) (This comparison was done by converting the costs from this study to USD in PPP terms per person.) This disparity is somewhat similar to the findings of Hutchings *et al.* (2017) who also found higher costs than those reported by WASHCost, and there are a number of possible explanations for this discrepancy. First, the schemes assessed in this investigation offered a higher service level than those included in WASHCost. In particular, all schemes served only private connections, whereas public taps are a common feature of rural piped schemes in India and sub-Saharan Africa. This is likely to have a major impact on both the upfront capital investment and recurrent expenditure when measured on a per capita basis. Clearly, it is likely to be more expensive to construct, operate and maintain a system that supplies 100 people through 20 private connections than through one public tap. In addition, most of the 14 schemes examined in this investigation had water treatment processes in place, which may not have been the case in the WASHCost systems. Another contributing factor could be the degree to which funds are directed towards routine, preventive maintenance rather than reactive CapManEx items. Notably, CapManEx reported by WASHCost schemes was considerably higher than schemes in this assessment, which may indicate that the WASHCost schemes were spending an insufficient amount on OpEx to mitigate costly equipment replacement (Nyarko *et al.* 2010). Further research is needed in the Vietnamese context to characterise how levels of OpEx influence CapManEx requirements in the long-term, including in relation to the relevant water supply service level.

CONCLUSIONS

This study sought to quantify the life-cycle costs associated with privately operated water services in rural Viet Nam. The results provide a valuable reference point for the development of cost norms (technical-economic standards), both in terms of the magnitude of costs and the drivers and distribution of those costs. Our study can inform the design of regulations for tariff formulation, support mechanisms such as subsidies and other incentives, initiatives to assist operators with business planning and data collection requirements and tools to aid these processes. We confirmed that even within a country, costs and 'cost norms' vary, and published studies demonstrate that this is also the case between countries. While life-cycle costs of water schemes are highly contextualised to the environment in which they exist, the process undertaken in this study to collect and analyse costs is highly relevant for, and transferable to, other countries and contexts.

The findings address an important knowledge gap in the Vietnamese rural water sector, and if responded to, would support Viet Nam to move towards achieving SDG target 6.1. Our study reveals the need for further nuanced and carefully contextualised LCCA, especially in combination with service delivery assessment, given that results show that both operational and capital expenditures varied widely. A range of factors influence unit costs, including scheme size, water source, climate and service level, while labour and electricity stand as the key operational cost drivers. Promisingly, most enterprises appeared to be financially viable, though further investigation is needed to determine whether tariffs are sufficient to cover future capital maintenance requirements and whether current levels of investment in capital maintenance are sufficient in light of capital depreciation.

It is important to stress that a good understanding of life-cycle costs is only one element of rural water sustainability and functionality. Conducting LCCA is an important step, but does not obviate the need to support other key building blocks related to service delivery and sector-level governance.

ACKNOWLEDGEMENTS

This study was funded by the Australian Department of Foreign Affairs and Trade (DFAT), though DFAT did not have involvement in the study design or in the collection, analysis and interpretation of data or writing of the report. The authors thank the research team who carried out this study including Doan Quang Huy Institute for Economics and Water Resources Management; Nguyen Hong Quan, Hanh Nguyen, Per Ljung and Thuy Bui Thanh from East Meets West Foundation, Viet Nam. The research team sincerely thank private enterprise owners and government representatives who were part of this study and who generously gave their time and shared their financial records and knowledge in support of the study.

CONFLICT OF INTEREST

None.

DATA AVAILABILITY STATEMENT

All relevant data are included in the paper or its Supplementary Information.

REFERENCES

Alexander, K. T., Mwaki, A., Adhiambo, D., Cheney-Coker, M., Muga, R. & Freeman, M. C. 2016 The life-cycle costs of school water, sanitation and hygiene access in Kenyan Primary Schools. *International Journal of Environmental Research and Public Health* **13** (7), 637.

Burr, P. & Fonseca, C. 2013 *Applying a Life-Cycle Costs Approach to Water Costs and Service Levels in Rural and Small Town Areas in Andhra Pradesh (India), Burkina Faso, Ghana and Mozambique*. IRC International Water and Sanitation Centre. Full Working Paper. Accessed from: https://www.ircwash.org/resources/washcost-working-paper-8-applying-life-cycle-costs-approach-water (accessed 15 September 2020).

Carrard, N., Madden, B., Chong, J., Grant, M., Nghiem, T., Ha, L. & Willetts, J. 2019 Are piped water services reaching poor households?: empirical evidence from rural Viet Nam. *Water Research* **153**, 239–250.

Centre for Rural Water Supply and Environmental Sanitation (CERWASS), Viet Nam Ministry of Agriculture and Rural Development 2000 *National Rural Clean Water Supply and Sanitation Strategy up to 2020*. Accessed from: https://www.ircwash.org/resources/national-rural-clean-water-supply-and-sanitation-strategy-2020 (accessed 15 September 2020).

Directorate of Water Resources, Viet Nam 2018 Report on rural water supply management. In: *Proceeding of Conference on Rural Water Supply Management in Bac Ninh Province Organized by MARD on 19 September 2018*.

Fonseca, C., Franceys, R., Batchelor, C., McIntyre, P., Klutse, A., Komives, K., Moriarty, P., Naafs, A., Nyarko, K., Pezon, C., Potter, A., Reddy, R. & Snehalatha, M. 2010 *Life-Cycle Costs Approach: Glossary and Cost Components. Briefing Note 1*. IRC International Water and Sanitation Centre. Accessed from: https://www.ircwash.org/sites/default/files/Fonseca-2010-Life.pdf (accessed 15 September 2020).

Fonseca, C., Franceys, R., Batchelor, C., McIntyre, P., Klutse, A., Komives, K., Moriarty, P., Naafs, A., Nyarko, K., Pezon, C., Potter, A., Reddy, R. & Snehalatha, M. 2011 *Life-Cycle Costs Approach: Costing Sustainable Services. Briefing Note 1a*. IRC International Water and Sanitation Centre. Accessed from: https://www.ircwash.org/sites/default/files/briefing_note_1a_-_life-cycle_cost_approach.pdf (accessed 15 September 2020).

Foster, T. & Hope, R. 2017 Evaluating waterpoint sustainability and access implications of revenue collection approaches in rural Kenya. *Water Resources Research* **53**, 1473–1490. doi:10.1002/2016WR019634.

Government of Viet Nam 2013 *Circular 45.Circular Guiding Regulation on Management, Use and Depreciation of Fixed Assets*. Ministry of Finance, Socialist Republic of Viet Nam. No. 45/2013/TT-BTC, Ha Noi, April 25, 2013.

Government of Viet Nam 2017 *National Action Plan for the Implementation of the 2030 Sustainable Development Agenda*. Accessed from: https://vietnam.un.org/index.php/en/4123-national-action-plan-implementation-2030-sustainable-development-agenda (accessed 15 September 2020).

Grant, M., Carrard, N., Madden, B., Willetts, J., Dominish, E., Bui, L. & Nghiem, T. 2016a *Access to Piped Water Services From Private Water Enterprises in Rural Viet Nam*. Enterprise in WASH – Research Report 7, Institute for Sustainable Futures, University of Technology Sydney.

Grant, M., Dominish, E., Carrard, N., Buy, L., Ha, H. & Nghiem, T. 2016b *Reducing or Increasing Inequalities? The Role of Private Water Enterprises in Rural Viet Nam, Development Bulletin, No 77*.

Hutchings, P., Franceys, R., Smits, S. & Mekala, S. 2017 *Community Management of Rural Water Supply: Case Studies of Success From India*. Oxon, Abingdon; New York, NY: Routledge, an imprint of the Taylor & Francis Group.

Hutton, G. & Varughese, M. 2016 *The Costs of Meeting the 2030 Sustainable Development Goal Targets on Drinking Water, Sanitation, and Hygiene World Bank*.

Jones, S. A., Anya, A., Stacey, N. & Weir, L. 2012 A life-cycle approach to improve the sustainability of rural water systems in resource-limited countries. *Challenges* **3** (2), 233–260.

Mansour, G. 2016 *Global Study on Sustainable Service Delivery Models for Rural Water – Country Brief*. World Bank, Vietnam.

Moriarty, P., Smits, S., Butterworth, J. & Franceys, R. 2013 Trends in rural water supply: towards a service delivery approach. *Water Alternatives* **6** (3), 329–349.

Nyarko, K. B., Dwumfour-Asare, B., Appiah-Effah, E. & Moriarty, P. 2010 Cost of delivering water services in rural areas and small towns in Ghana. In: *IRC Symposium 2010*. Pumps Pipes and Promises.

Reddy, V. R., Jayakumar, N., Venkataswamy, M. & Snehalatha, M. 2012 Life-cycle costs approach (LCCA) for sustainable water service delivery: a study in rural Andhra Pradesh, India. *Journal of Water, Sanitation and Hygiene for Development* **2** (4), 279–290.

UNICEF 2016 *Strengthening Enabling Environment for Water, Sanitation and Hygiene (WASH) Guidance Note*. United Nations Children's Fund (UNICEF), Water, Sanitation and Hygiene in Humanitarian Action. Accessed from: https://www.unicef.org/wash/files/WASH_guidance_note_draft_10_3_hr.pdf (accessed 15 September 2020).

UNICEF/WHO 2020 *Joint Monitoring Programme – Vietnam*. Available from: https://washdata.org/data#!/table?geo0=country&geo1=VNM (accessed 15 September 2020).

United Nations (UN) 2017 *Sustainable Development Goal 6: Ensure Availability and Sustainable Management of Water and Sanitation for All*. Accessed from: https://sustainabledevelopment.un.org/sdg6 (accessed 15 September 2020).

Vucijak, B., Grabus, A. & Avdic, S. 2015 *Tariff Setting Methodology for Water Supply and Sewerage Services in Bosnia and Herzegovina*. UNDP.

WASHCost 2012 *Providing a Basic Level of Water and Sanitation Services That Last: Cost Benchmarks, WASHCost Infosheet 1*. IRC.

Willetts, J., Grant, M., Carrard, N., Bui, L., Doan The, L., Pham Thi, D. & Dinh Van, D. 2017 Good water governance for inclusive growth and poverty reduction. In: *OzWater Conference*, Sydney. Available from: https://opus.lib.uts.edu.au/handle/10453/115352 (accessed 15 September 2020).

World Bank 2017 *Sustainability Assessment of Rural Water Service Delivery Models: Findings of a Multi-Country Review*. Washington, DC.

World Bank 2018a *World Bank Data – GDP Deflator – Vietnam*. Accessed from: https://data.worldbank.org/indicator/NY.GDP.DEFL.ZS?end=2016&locations=VN&start=1990 (accessed 15 September 2020).

World Bank 2018b *World Bank Data – Official Exchange Rate (LCU per US$, Period Average) – Vietnam 2016*. Accessed from: https://data.worldbank.org/indicator/PA.NUS.FCRF?end=2016&locations=VN&start=2016 (accessed 15 September 2020).

First received 21 February 2020; accepted in revised form 15 September 2020. Available online 5 October 2020

Research Paper

An assessment of penetration for pay-to-fetch water kiosks in rural Ghana using the Huff gravity model

Philip T. Deal and David A. Sabatini

ABSTRACT

Safe water enterprises across the developing world are attempting to meet demand for higher levels of water service. Existing, often free, water sources can make it difficult for these businesses to convince consumers to use a better-quality source or capture sufficient revenue for cost recovery. For this reason, it is imperative to develop a realistic understanding of penetration for small-scale water utilities. A cross-sectional assessment of 60 rural communities was used to evaluate the market share of a private service provider in Ghana. Household survey responses were used to identify the most attractive qualities of available water sources. Distance, taste, appearance, and affordability were found to be the most common motivational drivers. Using this information, a Huff gravity model was developed to assess the actual and potential market penetration and market share for the company in each community. The model and actual results agreed that about 38% of respondents would be regular customers at the given price. Even if water were free, the model predicted that the attractiveness of other sources would make it difficult to capture more than 58% of the sampled households. This illustrates the complexity of the water service ecosystem in a developing, rural context.

Key words | Huff gravity model, rural water supply, safe water enterprise, sub-Saharan Africa, willingness-to-pay

Philip T. Deal (corresponding author)
David A. Sabatini
School of Civil and Environmental Science,
University of Oklahoma,
202 W Boyd Street, Norman, OK 73019,
USA
E-mail: *philip.deal.125@gmail.com*

HIGHLIGHTS

- Novel application of the Huff gravity model on communal water sources to predict market penetration and market share of safe water enterprises.
- Unique dataset of household responses on top motivations for water source use.
- Penetration methodology verified by survey results, willingness-to-pay assessment, and market share evaluation.

INTRODUCTION

Safe water enterprises (SWEs) have become increasingly pervasive in water service provision across the world. SWEs are described as market-based approaches that deliver high-quality drinking water through decentralized solutions (Bhatnagar *et al.* 2017). Businesses vary in their treatment, transportation, delivery, and payment methods. Examples include home delivery of chlorinated water in India, ultrafiltration kiosks in Kenya, and private storefronts using reverse osmosis in Indonesia and Haiti (Cherunya *et al.* 2015; Patrick *et al.* 2017). These companies are

doi: 10.2166/washdev.2020.034

making efforts to increase levels of professionalism, institutional formality, and physical access to treated water in rural contexts (RWSN 2019).

However, the availability of treated water does not equate to usage. Market penetration rates historically range between 10 and 60% (Opryszko *et al.* 2013; Bhatnagar *et al.* 2017; Deal & Sabatini 2020). Previous studies have cited examples such as proximity, rainfall, or price as potential reasons for choosing between sources (Martínez-Santos 2017; Foster & Willetts 2018). This can lead to seasonal shifts in priority, periodic use of unimproved sources, or households choosing multiple sources for different domestic needs. Thus, it is imperative that market penetration be accurate in order to justify population coverage statistics and economic viability of SWEs. Due to pre-existing alternatives, it may be incorrect to assume that an SWE can achieve complete market saturation. More specifically, it can be hypothesized that an SWE, acting as a communal water provider, will be unable to reach a penetration rate approaching 100% as long as existing competition is readily available.

In this study, an assessment of rural households in Ghana investigated how motivational drivers relate to the uptake of SWE services. This article will first describe the study area and methodology of the research. Then, penetration rates at the district and community levels will be analyzed in relation to key respondent motivators. These results will culminate in a geospatial analysis of market demand using the Huff gravity model (Huff 1966), which estimates the probability that a given household will select each water source within their community. Finally, the effects of competition and water source attractiveness on rural water supply will be discussed.

STUDY AREA AND METHODS

Description of study area

The study occurred within in the Wassa East District of Ghana. In 2016, a baseline assessment estimated that the district had a population of around 90,000 people in 137 communities (Wassa East District Assembly 2016). The district is predominantly rural, with only three densely populated areas. It is within a tropical climate zone, with a mean annual rainfall of 1,500 mm. The total land area is about 1,652 km^2, but almost 600 km^2 is considered a forest reserve (GSS 2014). The district capital, Daboase, is about 38 km from Takoradi, the capital of the Western Region. About 69% of people claim commercial or subsistence farming as their primary employment, and nearly 80% of people do not have an education beyond Junior High School (Deal & Sabatini 2020). A map of the district, along with sampled communities, is shown in Supplementary Appendix A.

Access Development is a private, safe water enterprise in the district. The company extracts groundwater using solar-powered kiosks, which use microfiltration, UV disinfection, and chlorination for treatment. The company maintains their systems using circuit rider methods of operation. Area mechanics are stationed at various geographic locations across the district. They regularly visit each managed water source to provide maintenance and ensure continual water supply. Vendor communication and remote sensing assist in rapid response times.

The SWE is under a non-exclusive contract with the local government to work within the entire district, allowing for economies of scale. Customers collect water at the kiosk using the pay-to-fetch model. The price of water is set based on local market trends and efforts to cover operational costs and asset renewal. This is formally recorded in a signed agreement between the local government, each individual community, and the SWE. Other service providers in the area are predominantly community-based managers or religious institutions, but mining companies and public utilities exist in a few medium-sized towns as well. Deal & Sabatini (2020) provide further details on the institutional context and stakeholder exchanges.

Data collection and survey techniques

A cross-sectional study design (Jacobsen 2011) was utilized to collect quantitative data from households within the district. Two-stage cluster sampling was used to randomly select households from 30 intervention and thirty control communities. The data were used to determine the overall *market penetration rate*, which represents the proportion of households in intervention communities that choose to use SWE water at least weekly (Farris *et al.* 2010). Data collection was completed in 2019, after intervention communities had been under contract with AD for 6 months to 2 years. Water sources within each community were evaluated for reliability and quality. Testing included fecal coliforms, turbidity, conductivity, chlorine, and iron (Deal & Sabatini 2020).

Systematic random sampling of households occurred within each community. Enumerators received satellite maps, GPS coordinates, and sampling intervals to identify sampled households. Data were collected using the mWater platform. Unique mWater IDs were used to match household responses to previously identified water source characteristics and locations. Household heads or their spouse were given priority, though other adults were surveyed if necessary. GPS tracking allowed for the verification of survey completion. About 1,152 households were evaluated in total, and about 57% of respondents were female. Twenty-four households were removed through quality control checks of duplication, age, inconsistency, improper mental states, or poor accuracy ratings (Deal & Sabatini 2020).

Informed consent was required prior to participating in the study. Identifying information was removed for the subjects' protection. The research protocol was cleared by the University of Oklahoma IRB. Data collection was carried out by local enumerators after completing training for ethical, in-person interviews. During preliminary survey testing, questions were asked in English, Fante, and Ewe after translation into local dialects.

Motivational options were presented from a pre-determined list derived from historical sources (Cherunya *et al.* 2015; Martínez-Santos 2017; Foster & Willetts 2018), both from positive and negative perspectives (Supplementary Appendix B). When open-ended responses occurred, they were reallocated to their respective categories. The statistical analysis of respondent data was completed using IBM SPSS Statistics 26.

Data analysis

Actual penetration results were compared with the market demand produced by the Huff gravity model (Huff 1966). This competitive location model combines the variables of distance, facility attractiveness, and demand to predict the market share that a given store location can capture. It has historically been used to identify optimal locations for grocery stores, gas stations, shopping centers, or other retail outlets. While any environment with an extensive piped water system might not find this tool appropriate, many rural and peri-urban areas still use communal water sources where these predictions could be valuable.

In this study, Equation (1) defines the market penetration, or the probability (P_{ij}) that a given household located at given location, i, would choose to collect water from source, j. This assumes that A_j is a measure of the attractiveness of each water source, and D_{ij} is the distance between the household and the water source. Euclidean distance was calculated using Equation (2), which is based on the radius of the earth, r, the latitude, ϕ, and the longitude, λ. The Multiplicative Competitive Interaction (MCI) model for attractiveness was used to test multiple factors, as shown in Equation (3) (Cooper & Nakanishi 1988). The MCI model is an econometric model for analyzing market shares. The variables, α and β, act as empirical model parameters, where the default α value is assumed to be 1, and β falls between -1 and -2 (Dolega *et al.* 2016). Then, the model (Equation (1)) was transformed into a linear form using Equation (4). This allowed for linear regression analysis and optimization of the model. The variables, Π_i, Λ_j, and Δ_i, are the geometric means of P_{ij}, A_{jh}, and D_{ij}, respectfully.

After the market penetration of SWE water (P_{SWE}) is estimated, the theoretical results can be compared with the market share (M_S) derived from respondent data. Market share, shown in Equation (5), represents the percentage of an industry's total sales that is earned by a particular company over a specified time period (Farris *et al.* 2010). Actual results are derived using consumer data for the mean spending on SWE water ($\bar{S}_{\text{SWE}}$) and total water ($\bar{S}_T$), which also represents potential buying power.

$$P_{ij} = \frac{A_j^{\alpha} D_{ij}^{-\beta}}{\sum_{j=1}^{n} A_j^{\alpha} D_{ij}^{-\beta}} \tag{1}$$

$$D_{ij} = 2r * \arcsin\left(\sqrt{\sin^2\frac{\varphi_i - \varphi_j}{2} + \cos\varphi_j \cos\varphi_i \sin^2\frac{\lambda_i - \lambda_j}{2}}\right) \tag{2}$$

$$A_j^{\alpha} = A_{j1}^{\alpha_1} * A_{j2}^{\alpha_2} * \cdots * A_{jh}^{\alpha_h} \tag{3}$$

$$\log\left(\frac{P_{ij}}{\Pi_i}\right) = \sum_{h=1}^{H} \alpha_h \log\left(\frac{A_{hj}}{\Lambda_j}\right) - \beta \log\left(\frac{D_{ij}}{\Delta_i}\right) \tag{4}$$

$$M_s = \frac{\bar{S}_{\text{SWE}}}{\bar{S}_T} = \frac{P_{\text{SWE}} \bar{S}_T}{\bar{S}_T} = P_{\text{SWE}} \tag{5}$$

RESULTS AND DISCUSSION

Descriptive analysis

Intervention communities had, on average, a choice of about 6 communal water sources, with a range from 2 to 14 options. Each community had a relatively new (6 months to 2 years) SWE water source available and functional, specifically 52 of the 177 identified sources. The SWE sources included 48 kiosks with vendors selling water using pay-to-fetch and four boreholes under a monthly tariff scheme. Community penetration rates for SWE sources tended to increase as available alternatives decreased, emphasizing the importance of local options (see Supplementary Appendix C). The control communities had a total of 145 water sources with a statistically similar distribution of surface, groundwater, and piped sources. When surveyed, households were asked what motivated them to choose or *not* choose a given primary source, with results summarized in Figure 1.

From the positive perspective (Figure 1(a)), top motivations included close proximity (distance), appearance, and taste. Intervention households saw increased proportions of good quality and affordability determinants. When households were asked what motivated them *not* to choose an alternative drinking water source (Figure 1(b)), distance and taste were still relevant. However, never being able to afford or only sometimes being able to afford had noticeable spikes in intervention households. This information provides evidence for a sub-section of the population that feels they cannot afford to regularly pay for water, thus limiting the market in a rural context. Further delineations between specific customer classifications are illustrated in Supplementary Appendix D.

It is also important to note that the community penetration rate of SWE water was statistically correlated with negative motivations of poor taste ($p = 0.042$) and, separately, not having the ability to pay ($p = 0.049$). Essentially, an individual community cluster may have a spike of complaints for either factor, which suggests an underlying cause for the lower local uptake observed. Upon further investigation, four communities with greater affordability complaints were observed to resist patronage of kiosks in an effort to lower pay-to-fetch costs. As another example, when isolating communities with elevated chemicals related to poor taste ($n = 14/30$), a significantly lower mean penetration rate was observed (30.4 versus 57.0%; $p = 0.013$). Tested chemicals included chlorine, conductivity, and iron (WHO 2017; Deal & Sabatini 2020). Quality (*E. coli*) and appearance (turbidity) indicators were also examined, but no significant trends with penetration were observed.

Finally, Figure 2 summarizes the consumer data in the dry season for spending on total water, sachet water, and SWE water per week according to each customer class (see Supplementary Appendix D). Also, collected volumes from a household's primary source are compared with the

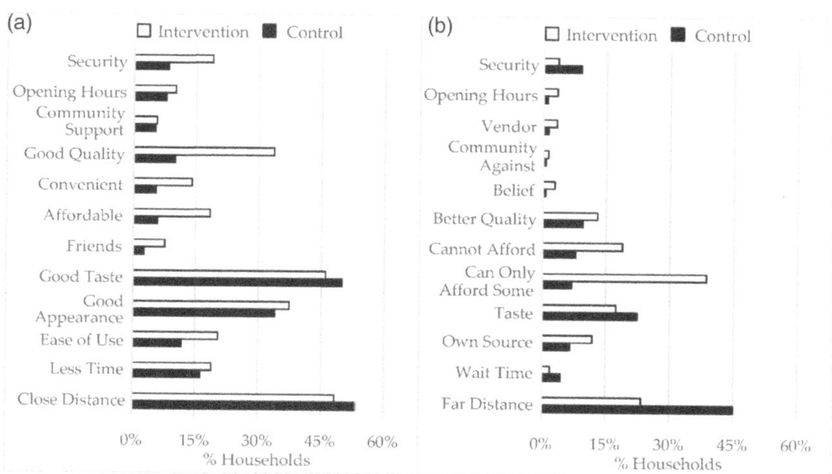

Figure 1 | Motivations to (a) choose or (b) NOT choose a given primary source.

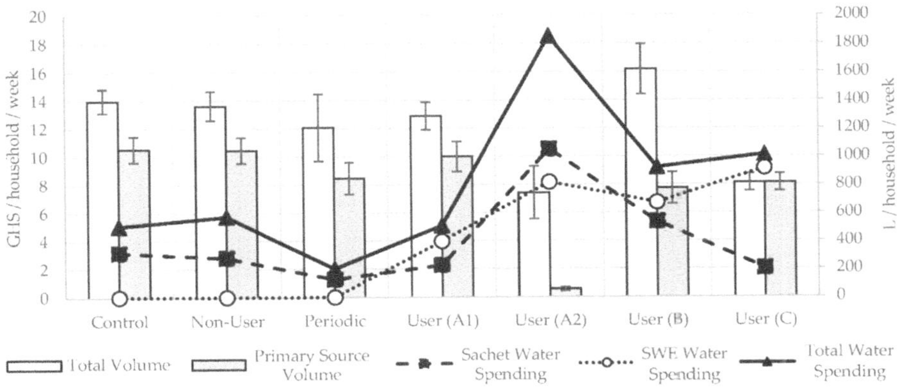

Figure 2 | Household water volume collected and water spending per week by customer classification.

total volume collected each week. The differences observed represent a portion of the sample population that uses multiple sources to meet their domestic needs (Deal & Sabatini 2020). In the intervention group, mean total spending on water was 7.07 GHS per household per week and mean total volume collected was 1,345 L per household per week.

Huff gravity model

The descriptive data analysis provided the necessary information to develop a Huff gravity model for each community in the district. The key assumption of this model is that each consumer is influenced by their distance from a given retail location and the attractiveness of the source. Table 1 summarizes the variables used within the model, and Figure 3 shows the progression of model development for an example community.

In Figure 3(a), it can be observed that there are five water sources within the community, each functional at the time of the survey. For reference, WP1 and WP3 are unprotected hand dug wells, WP2 is a protected hand dug well, WP4 is an SWE kiosk, and WP5 is a collection point along a stream. Randomly selected household sample locations are also noted. Figure 3(b) represents the highest probability value predicted by the model for all water sources, and a push-pull effect can be observed. This is the product of the attractiveness and distance for each water point in Equation (1). While attractiveness is static according to the model variables chosen, distance inversely reduces attraction.

Table 1 | Motivational variables associated with penetration and their Huff model scoring limits

			Scoring limits	
Motivation	**Variable**	**Type**	**Low**	**High**
Availability	Functionality	Binary	Not functional	Functional
	Annual reliability	Ordinal	<6 months	12 months
Taste (Chem.)	Chlorine	Binary	>0.3 mg/L	<0.3 mg/L
	Conductivity	Binary	>300 μS/cm	<300 μS/cm
	Iron	Binary	>0.3 mg/L	<0.3 mg/L
Appearance (Chem.)	Turbidity	Binary	>5 NTU	<5 NTU
Quality (Bio.)	Free coliforms	Ordinal	>100 MPN/100 mL	0 MPN/100 mL
Organoleptic rank	Taste, odor, appearance, lather	Ordinal	Lowest rated choice	Highest rated choice
Affordability	Price	Ordinal	0.30 GHS/18 L	Free
Social pressure	Community resistance	Binary	Observed	Not observed

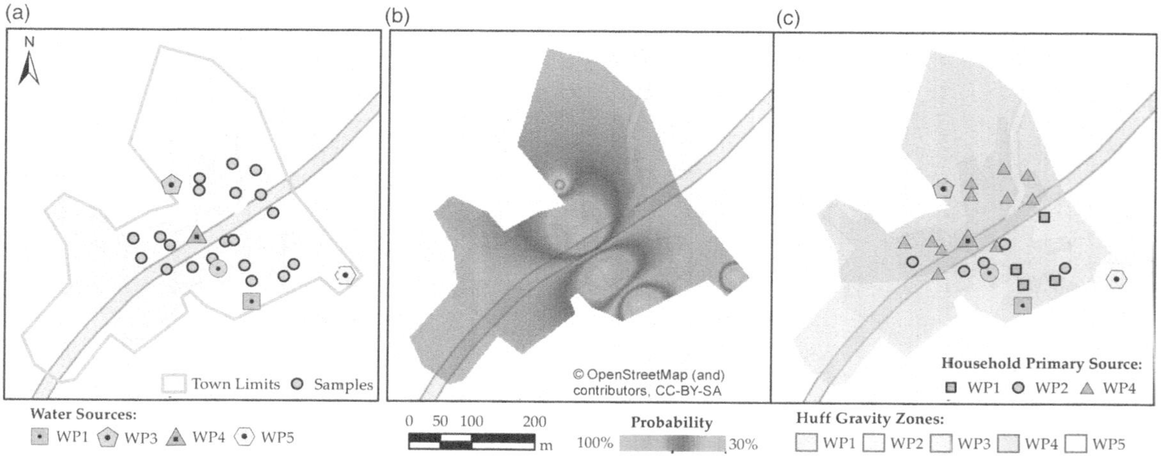

Figure 3 | Huff model analysis: (a) sample distribution for an example community, (b) max probability distribution for all sources (gravity map), and (c) highest probable water source versus respondent's primary source.

Figure 3(c) maps out each source's radius of influence, as well as the actual primary drinking water source for each household. The spatial influence is clearly evident, supporting the motivation of close proximity for many households. While a few households appear to be incorrectly assigned to a particular source, it is important to use both Figure 3(b) and 3(c). Some sources far away only have around a 30% chance of using the predicted source, meaning that the others are offering considerable competition.

Attractiveness was calculated by incrementally adding the variables identified in Table 2. These variables were determined from the descriptive analysis and historical determinants (Martínez-Santos 2017). They were restricted to traits of the water source independent of household traits, such as socioeconomic status or size, in order to simulate a new installation decision for a safe water enterprise. Some determinants, such as reliability, quality, and taste rank, were considered ordinal, while other determinants were considered binary, such as functionality and social pressure. While ordinal variables offered incremental improvements to attraction, binary variables could attribute a harsh penalty if a condition was violated (see Table 1). Each was optimized based on the best performance of the model using multinomial linear regression in the log-normal form of the probability function (Equation (4)). Table 2 shows the details of the model progression.

From Table 2, gradual improvements are observed from Model I to VII with respect to the regression fit. Distance was consistently a significant determinant. However, reliability and functionality seemed to have some redundancy due to their dependent nature. The penalty for non-functionality ($A_j = 0.01$) serves the important role of ruling out unavailable water sources, but must be non-zero due to the log transformation. The chemical taste determinant, which applied a penalty to sources with high conductivity or iron, was found to be more significant as other factors were applied. For this reason, it was kept instead of the taste rank factor. The opposite decision was used for appearance, where turbidity was generally found to be counterproductive and appearance rank performed better. The effect of the rank variables, as shown in model IV, was similar for all organoleptic properties, as they seemed to highlight the favored water sources in a given community. Quality was also found to be counterproductive, suggesting that the majority of households do not strongly consider it. This is not surprising, given that public water quality testing in the district is rather limited. Social pressure and price proved to be rather important, but for different reasons. When a penalty was applied to sources that were observed to have negative social pressures, such as a borehole with evidence of worms or attempts to reduce kiosk prices, a significant jump in correlation occurred for model VI. While a price factor did not have a significant effect on the linear regression

Table 2 | Results of multiple-regression analysis for different models of probability; modeled determinants of water source usage are shown, and the subsequent penetration rates for SWE water sources are included

Model predictions						**Variable *p*-values**[c]											
Models	**Penetration (%)**[a]	**Adj. R^2**	**Ln (Adj. R^2)**	**F-value**	**F-sig.**[b]	**Dist.**	**Func.**	**Reliability**	**Taste (Chem.)**	**Turbidity**	**Quality**	**Taste rank**	**Odor rank**	**Appear. rank**	**Lather rank**	**Social**	**Price**
I	45.7	0.279	0.018	123.9	<0.001	0.009											
II	53.3	0.526	0.325	354.2	<0.001	0.001	0.140	0.001	0.260								
III	68.6	0.468	0.343	280.8	<0.001	0.009	0.268	0.003	0.246	0.469	0.001						
IV	57.0	0.600	0.738	478.0	<0.001	<0.001	<0.001					<0.001					
V	58.3	0.640	0.769	566.4	<0.001	<0.001	0.023	0.396	0.008				<0.001	<0.001	0.010		
VI	45.3	0.740	0.776	906.7	<0.001	<0.001	0.017	0.356	0.020				<0.001	0.001	0.008	0.001	
VII	38.8	0.725	0.775	839.3	<0.001	<0.001	0.017	0.363	0.021				<0.001	0.001	0.008	0.001	0.951

[a]Model prediction of % households using SWE water services.
[b]Confidence level 95%; F-sig = 0.05.
[c]Confidence level 95%; $p = 0.05$.

($p = 0.951$), it dropped the penetration rate into the appropriate range. Given its good correlation with actual results, model VII was selected as the optimal choice. The comparison between actual and predicted probability values is shown in Figure 4.

With the market penetration rate of 38.8% derived from model VII, the theoretical and self-reported market values could be compared. The actual market share was 35.2%, which was based on the respondent mean spending on SWE water of 2.49 GHS per household per week. Assuming the mean spending on total water was 7.07 GHS, the model predicts the average household would spend 2.74 GHS each week on SWE water (Equation (5)), which was statistically similar to the actual results ($p = 0.454$).

While providing useful insights, each model has the following limitations. Tested determinants are not exhaustive, with factors such as age or flow rate potentially increasing accuracy. As most communities were rural in this study, network analysis was deemed unnecessary. This accounts for increased distance due to movement along street grids, but only a few peri-urban environments could have seen some benefits. Also, a few instances of error were observed where water sources were near schools or markets, and it is assumed respondents living far away may collect on the

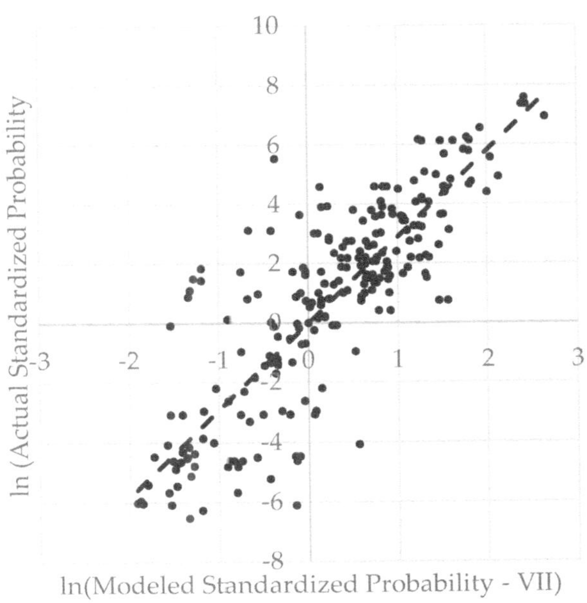

Figure 4 | Predicted versus actual probabilities for using a given water source using the log-normal standardized probability.

way home from work or school. Treating streams and other surface sources as lines or polygons rather than collection points could also impact model results. Due to the cluster sampling methodology, the assumption was made that mean water spending was representative of buying power for the district, but it should ideally be community specific (Drezner & Drezner 2004). Finally, this model will change with present day market structure and trends, such as new laws or companies. The temporal study conducted by Opryszko *et al.* (2013) showed that penetration can change over time, emphasizing the importance of recently collected data in future iterations.

Two additional limitations should be pursued in future research. First, considering fluctuations in facility attractiveness due to rainfall patterns would better represent seasonal priorities (Hope *et al.* 2020). In this region of Ghana, many households would set out large barrels to collect rain, essentially creating an onsite source of water based on rainfall patterns. Second, the use of multiple sources impacts purchase volumes and should be incorporated. In this study, only 20.1% of total water collected is from an SWE source, despite capturing 35.2% of the market share of revenue.

Discussion

The results presented in this study provide evidence to support the hypothesis that market saturation, or approaching 100% penetration, is unrealistic for a safe water enterprise using communal water sources. The Huff gravity model illustrates how existing alternatives still have attractiveness, either by being closer to the home, a cheaper price, or better tasting. These effects will most likely decrease as household piped connections become more prevalent, but are very applicable when central collection points exist.

For the Wassa East district, the actual SWE penetration rate within the intervention group was determined to be 37.8, 95% CI [31.4, 44.7], in a companion study (Deal & Sabatini 2020). The maximum penetration predicted in the various Huff models was 68.6%, which assumed that distance, reliability, and quality were the most valued traits of a water source. Understanding that taste and appearance are more valuable to the average household, it is more reasonable to assume that the maximum penetration attainable *with existing competition* is roughly 58% (model V), without considering affordability. With all factors considered while holding the market rules constant, the Huff model predicted that penetration would reduce to 38.8% (model VII). This falls within the confidence interval of the household survey results and is corroborated by a concurrent willingness-to-pay assessment summarized in Supplementary Appendix E. The market share of available revenue (35.2%) also reflected the market penetration rate.

Affordability was a critical motivator for certain households within the intervention population. While this did not play a significant role in the control group (6–8%), 18.9% of intervention households claimed they could never afford SWE water regularly. Further, if mean spending on total water for the district is 7.07 GHS per week, a household can afford about 636 L from the SWE at its current price. This was sufficient for drinking, cooking, and some hygiene needs, but insufficient for bathing and cleaning volumes reported by the control group. This suggests an absolute limit for market penetration at the current price and emphasizes the importance of maintaining cheaper, but safe, alternatives in this context. Other improved sources likely allow for the use of multiple sources and take away from the financial sustainability of the SWE, and vice versa. However, their availability also provides leverage to keep the private company accountable in both price and performance of the service.

It was interesting to observe the market attempt to influence the price of water, which was captured by the 'social pressure' factor in the Huff model. For communities that wanted a price drop from 0.20 to 0.10 GHS/ 18 L, a 'boycott' occurred. The higher price, which is common for standpipes in a few of the medium-sized towns, was agreed upon in a signed agreement prior to construction. However, the community applied pressure to renegotiate to common handpump prices (0.10 GHS/ 18 L) after the capital investment had already occurred. In this circumstance, pressure was possible because of the existing alternatives available and because the local government had allowed their continued use. A change in regulation or contract terms, such as exclusivity rights, delegated management areas, or pro-poor legislation, could theoretically change these conditions.

However, the motivations behind the Huff model suggest that competition will continue to be a factor, regardless of formality, and should not be dismissed.

CONCLUSIONS

Overall, the Huff gravity model (model VII) and the household survey both agreed that the actual penetration in the Wassa East District should fall between 35 and 40% at the time of the study. This accounted for the attractiveness of individual water sources, the distance from each household, and the differing prices. For similar SWEs, this predictive model can be used to estimate market share based on company decisions on location and price. With only a water point inventory of a given area, it can help to determine the proportion of the population actually served, rather than assuming an entire community is benefiting. With the addition of consumer data, these results can be used as part of a financial projection tool to predict cost recovery for the targeted district. Practically, Access Development has applied this modeling technique in this capacity.

Formal SWEs are looking to fill gaps in water service provision in urban and rural communities across the globe. Many examples offer treated, improved water and reliable service to their customers in an effort to provide continuous supply and protective health benefits. Their goal is to offer a better product in scenarios where existing piped, communal, and unsafe water alternatives are already providing for the population. This study has shown that an SWE will inherently face reduced penetration due to competition and differing motivations of its customer base. It is critical that implementers, funders, and policy-makers understand these implications, allowing for effective institutional arrangements and a realistic perspective of the role that these businesses can fill in the water provision ecosystem.

DATA AVAILABILITY STATEMENT

All relevant data are available from an online repository or repositories. Visit: http://dx.doi.org/10.17632/b6gs9fd33z.2.

REFERENCES

Bhatnagar, N., Lampert, S., Goyal, V., Mehta, R., Rai, R. & Chandrawekhar, A. 2017 *The Untapped Potential of Decentralized Solutions to Provide Safe, Sustainable Drinking Water at Large Scale: The State of the Safe Water Enterprises Market*. Washington, DC.

Cherunya, P. C., Janezic, C. & Leuchner, M. 2015 Sustainable supply of safe drinking water for underserved households in Kenya: investigating the viability of decentralized solutions. *Water (Switzerland)* 7 (10), 5437–5457. https://doi.org/10.3390/w7105437.

Cooper, L. G. & Nakanishi, M. 1988 *Market Share Analysis: Evaluating Competitive Marketing Effectiveness*. Kluwer Academic Publishers, Boston.

Deal, P. & Sabatini, D. 2020 Evaluating the level of household water service provided by a private water enterprise in Ghana. *Water* **12** (3), 693.

Dolega, L., Pavlis, M. & Singleton, A. 2016 Estimating attractiveness, hierarchy and catchment area extents for a national set of retail centre agglomerations. *Journal of Retailing and Consumer Services* **28**, 78–90. https://doi.org/10.1016/j.jretconser.2015.08.013.

Drezner, T. & Drezner, Z. 2004 Finding the optimal solution to the Huff based competitive location model. *Computational Management Science* **1** (2), 193–208.

Farris, P. W., Bendle, N., Pfeifer, P. E. & Reibstein, D. 2010 *Marketing Metrics: The Definitive Guide to Measuring Marketing Performance*. Pearson Education, Upper Saddle River, NJ.

Foster, T. & Willetts, J. 2018 Multiple water source use in rural Vanuatu: are households choosing the safest option for drinking? *International Journal of Environmental Health Research* **28** (6), 579–589. https://doi.org/10.1080/09603123.2018.1491953.

Ghana Statistical Service (GSS). 2014 *Wassa East District – 2010 Population & Housing Census – District Analytical Report*.

Hope, R., Thomson, P., Koehler, J. & Foster, T. 2020 Rethinking the economics of rural water in Africa. *Oxford Review of Economic Policy* **36** (1), 171–190.

Huff, D. L. 1966 A programmed solution for approximating an optimum retail location. *Land Economics* **42** (3), 293. https://doi.org/10.2307/3145346.

Jacobsen, K. H. 2011 *Introduction to Health Research Methods*. Jones & Bartlett Publishers, Burlington, MA.

Martínez-Santos, P. 2017 Determinants for water consumption from improved sources in rural villages of southern Mali. *Applied Geography* **85**, 113–125. https://doi.org/10.1016/j.apgeog.2017.06.006.

Opryszko, M. C., Guo, Y., MacDonald, L., MacDonald, L., Kiihl, S. & Schwab, K. J. 2013 Impact of water-vending kiosks and hygiene education on household drinking water quality in rural Ghana. *American Journal of Tropical Medicine and*

Hygiene **88** (4), 651–660. https://doi.org/10.4269/ajtmh.12-0065.

Patrick, M., Steenland, M., Dismer, A., Pierre-Louis, J., Murphy, J. L., Kahler, A., Mull, B., Etheart, M. D., Rossignol, E., Boncy, J., Hill, V. & Handzel, T. 2017 Assessment of drinking water sold from private sector kiosks in post-earthquake Port-au-Prince, Haiti. *American Journal of Tropical Medicine and Hygiene* **97** (Suppl 4), 84–91. https://doi.org/10.4269/ajtmh.16-0692.

Rural Water Supply Network (RWSN) 2019 Chapter 3: The Directory In: *The 2019 RWSN Directory of Rural Water Supply Services, Tariffs, Management Models and Lifecycle Costs* (P. Deal & S. Furey eds.). Skat Foundation, St. Gallen Switzerland.

Wassa East District Assembly 2016 Personal Communication.

WHO 2017 *Guidelines for Drinking-Water Quality: First Addendum to the Fourth Edition*. World Health Organization, Geneva.

First received 15 February 2020; accepted in revised form 30 June 2020. Available online 22 October 2020

Research Paper

Understanding water demand and usage in Mandalay city, Myanmar as a basis for resetting tariffs†

Tanvi Nagpal, Henry Rawlings and Maël Balac

ABSTRACT

Water service providers must understand the needs and consumption patterns of their customers in order to make informed decisions on new investments. The Mandalay City Development Committee is undertaking an expansion of its water supply and sewerage infrastructure with a goal of reaching every citizen. The service provider needs to revise tariffs to cover higher provision costs. To this end, a survey conducted in 2019 in Mandalay City, Myanmar revealed that 80% of users relied on more than one water source. The selection of water source depended on intended use, quality, availability, and reliability more than it did on price. Users are heavily reliant on groundwater, which is also increasingly scarce and contaminated. The findings shed light on the challenges of meeting operational costs through tariffs when municipal supplies are not the sole source of water, water quality is poor, and non-revenue water is high.

Key words | groundwater, Myanmar, piped water supply, service delivery, tariff setting, water demand

Tanvi Nagpal (corresponding author)
Henry Rawlings
Johns Hopkins University School of Advanced International Studies,
1717 Massachusetts Ave.,
Washington, DC 20036,
USA
E-mail: *tnagpal1@jhu.edu*

Maël Balac
IPE Triple Line,
London,
UK

†This study conformed to Johns Hopkins University IRB procedures, and received the status of Exempt on March 27, 2019. The study number is HIRB00008788. Further documentation is available if necessary.

HIGHLIGHTS

- Primary data on water supply in Myanmar are relatively scant. We believe that this is the only primary data collected in Mandalay that is representative of the city's water demand and usage.
- The research provides evidence to support the hypothesis that in low-income settings, tariff setting is complicated by the presence of multiple, substitutable water sources.
- The study also presents surprising evidence that quality and reliability may matter more to a low-income urban population in their choice of drinking water solutions than does price.
- The study makes data-based short- and medium-term policy recommendations based.

INTRODUCTION

While there has been undeniable progress in access to improved sources of water, almost a third of the world's population still lacks access to a reliable source of safe water (WWAP 2019). Much of this underserved population lives in rural areas; however, rapid urbanization, particularly in middle- and low-income countries has left many governments struggling to meet urban demand. Global access to safe water is highly correlated to income, and both rural and urban-poor populations suffer disproportionately (WWAP 2019). Myanmar is no exception. Almost 70% of the country still resides in rural areas, where 22% of citizens lack access to safe water compared to 6% in urban areas (CSO 2018). Myanmar will urbanize extensively over the next decades. One third of rural residents are expected to move to cities in the next 20 years, and half of the country's population will live in urban areas by 2050

doi: 10.2166/washdev.2020.076

(UNDESA 2018). Migration and natural population growth will test urban infrastructure and present an important challenge to city managers overseeing public service provision.

Water scarcity and climate change will exacerbate the challenge of rapid urbanization. By 2040, the number of people living in settings of extreme water stress will increase by an estimated 20% around the world (UNICEF 2017) (UNICEF defines extremely high levels of 'water stress' as those that exceed a ratio of 0.8 water use to supply, indicating a very high level of competition for available water (UNICEF 2017).). The combination of climate change and economic growth means that already arid regions, such as Mandalay, will not only be drier but also have substantially higher water needs. Myanmar has ranked among the most climate-vulnerable countries for the past two decades (Eckstein *et al.* 2019). Across the country, regions have observed shorter monsoon periods, longer pre-monsoon droughts, and higher rainfall depths (Tham 2010). Other studies have identified water scarcity, infrastructure, and quality control as barriers to water security for all citizens (Pink 2016). As leaders brace for the coming decades of changing environments and demographic shifts, understanding consumption patterns and usage has become critical for governments around the world.

Little was known about public service provision in Myanmar until the early part of the 2000s. In 2011, the government undertook a series of public sector reforms including increased transparency, decentralized governance, and socio-economic development programs (Rab *et al.* 2016). The international community responded by lifting or suspending economic sanctions and by providing 'extensive international assistance for Myanmar, including for its urban development and water sector' (ADB 2017a). Among the international financial and technical assistance were projects to improve urban water supply in Mandalay City – Myanmar's second-largest city and former royal capital. In 2016, Mandalay committed to borrow $100 million from the Asia Development Bank and the French Development Agency (AFD) to improve municipal sewage and expand piped water supply (ADB 2016). Other organizations, like the Japan International Cooperation Agency (JICA), GRET, and Vides-Evides International (VEI), are also supporting development projects in Mandalay City. However, an extended humanitarian crisis in the Rakhine State has threatened the flow of foreign investment and donor financing since 2018, placing further emphasis on domestic resource mobilization (IMF 2019).

With over 1.3 million people within its jurisdiction, Mandalay City is considered an economic and cultural hub of Upper Myanmar. Mandalay's population has grown at a rate of 10% per year, faster than Myanmar's total population rate of 1.8% (DOP 2015; ADB 2017b). The Mandalay City Development Committee (MCDC) manages the city, with two departments that oversee water-related issues. One is responsible for infrastructure installation, operations, and maintenance. A second, which falls under the Revenue Department, deals solely with water revenue. Under *The City of Mandalay Development Law*, MCDC is responsible for implementing water supply works and 'determining, revising, assessing and collecting taxes and duties and rates thereof in respect of development works' (State Peace and Development Council Law No. 8/2002). Water tariffs represent one of the highest sources of revenue for the city, while water supply is also one of the highest costs. Although MCDC has improved many aspects of its piped water service, significant challenges remain. According to available estimates, MCDC piped water supply only covers about 55% of the population, tariffs do not cover the cost of limited water provision and reported non-revenue water ranges from 50 to 75% (ADB 2017b). The next section briefly looks at approaches to understanding water demand. A description of the survey methodology, findings, and short discussion follows.

URBAN WATER DEMAND

The present study looks at the intersection of urban water demand patterns and tariff setting in developing economies and contributes to the research specific to Myanmar (The present study was part of a recent International Growth Centre project conducted for the MCDC to inform policy decisions on service expansion.). Budgetary and time constraints prevented us from completing a full willingness to pay (WTP) analysis. However, the survey highlighted the main challenges MCDC faces as it considers how to finance operations and improvements to its water infrastructure and management.

There is a wealth of literature on estimating effective demand and tariff setting for water services. Arbués *et al.* (2003) and Worthington & Hoffman (2006) provide valuable insights into modeling price elasticities and variable choice considerations. Whittington & Nauges (2010) study estimation techniques in cities which face unique challenges, like the persistence of unmetered consumption, unreliable usage data, as well as the prevalence of multiple water sources. They find that modeling demand for water in these settings requires data from the alternative sources, which are often unavailable or hidden in informal markets. Other studies focus on the determinants of consumption, but they are largely clustered in rural areas (Basani *et al.* 2008; Basu *et al.* 2017; Martinez-Santos 2017; Gross and Elshiewy 2019; Wagner *et al.* 2019). In these settings, distance, perceived quality, and price influence the choice of water source.

Another section of the literature analyzes the response to changes in price or quality of water supply. In general, studies find that demand for piped water in urban areas is quite inelastic when source substitution is difficult (Nauges and Whittington 2010; Marzano *et al.* 2018). Importantly, when piped water and non-piped water are substitutes, households with access to multiple water sources (such as water venders, hand-dug, or deep wells) may be more sensitive to price changes than those relying solely on piped water, switching to cheaper sources when tariffs or prices are raised (Nauges and Whittington 2010).

WTP is positively correlated to incomes, ease of access, and reliability – especially in urban areas, and consumers will pay more for improved services (Casey *et al.* 2006, Orgill *et al.* 2013). Households tend to be more willing to pay for private water connections than public ones (North & Griffin 1993). Households without access to reliable and safe piped water engage in many 'coping behaviors'. One study finds that such coping costs can be double a monthly water bill but are still lower than WTP for improved water supply (Pattanayak *et al.* 2005). Furthermore, WTP for proposed improvements in water access declines as a household's baseline access to other water services improves (Van Houtven *et al.* 2017).

Tariff setting has been explored extensively as well. Local authorities typically hold responsibility for raising revenue sufficient to cover operational costs, ideally through a combination of local taxes (income, property, and sales tax for example), tariffs, and fees. In emerging economies, however, most utility providers are unable to cover even ongoing maintenance costs, due to large system losses from breakages and theft, poor control over investment and tariff setting decisions, and rapid urbanization (Nagpal *et al.* 2019). Recent research suggests that only 35% of utilities in developing countries are able to cover operations and maintenance costs (Lombana Cordoba *et al.* 2019).

User fees take on various forms, such as flat fees, volumetric or increasing block tariffs, and they may also reflect environmental costs and equity goals (Silva Pinto & Cunha Marques 2015). But the setting of these tariffs is often a political exercise and not based on financial, environmental, or social costs.

In many cases, the poorest households cannot connect to the piped water system despite the premise that tariffs should be kept low for social equity concerns. This challenge stems from the high upfront costs of connecting and the pervasiveness of irregular incomes in informal employment (Mitlin & Walnycki 2019). Thus, low tariffs end up being regressive as they subsidize the consumption of relatively wealthier households who can afford the connection costs and are able to pay a regular bill (van den Berg & Danilenko 2017). Under these conditions, poorer households rely on a range of sources to satisfy their household needs, using a combination of open, unimproved sources, such as ponds and lakes and safe sources, such as deep wells or purchased bottled water.

The following section briefly describes the study area and survey methodology.

METHODOLOGY

We combined in-depth interviews (IDIs) with a quantitative survey of households and micro, small, and medium enterprises (MSMEs) to draw conclusions about household and business demand and consumption patterns in the city's six townships.

Sample site: Mandalay city

Mandalay District is located in the Central Dry Zone of Myanmar and sits on the Irrawaddy River – the largest river in the country. With over 1.3 million people within its jurisdiction, Mandalay City is considered an economic and cultural hub of Upper Myanmar. MCDC governs the jurisdiction, which comprises of seven townships. The township populations range from 197,175 citizens in Chanayetharzan to 283,781 in Chanmyatharzi. Of the seven townships, only Amarapura and Patheingyi are not fully urban, with only 34 and 5% urban population, respectively (DOP 2015). In 2018, an average household had a monthly income of 350–450,000 MMK or 245–315 USD (The figures reported in dollars use a 1 Myanmar Kyat (MMK) = 0.0007 United States Dollar (USD), which is an average of the exchange rate conversion for 2018), and comprised of 4.8 members (Ford *et al.* 2019). In the city's jurisdiction, 81% of households own or rent their dwelling (DOP 2015).

Quantitative survey

The quantitative survey was designed to investigate water demand and to capture behavioral aspects, such as perception, satisfaction, and expectations from piped water supply. The survey spanned six urban townships (including Amarapura). Due to the absence of credible data on the actual proportion of businesses per household at the township level, it was assumed to be proportional to the ratio at the city level. Therefore, as per Probability Proportional to Size sampling, the number of surveys administered followed an 80% households (HHs) to 20% businesses (MSMEs) ratio (Lavrakas 2008; Skinner 2016). Table 1 provides a comparison between the initial sampling plan and the population captured by the actual survey.

A total of 1,480 respondents were randomly selected across the 6 townships, including 1,178 (80%) households and 30 (20%) MSMEs, distributed proportionally to the actual population. Eighteen enumerators were identified based on educational qualification, sector experience, and geography. The survey tool was pretested, and enumerators were trained over an interactive four-day training workshop in Mandalay. The map of the survey sample is displayed in Figure 1.

IDIs were conducted with senior officials to understand their perceptions of current MCDC operations and costs. Survey results were shared with MCDC officials and the Mayor's office.

RESULTS

The household survey and key informant interviews found that most residents in Mandalay City use many sources of water. Households and MSMEs rely on piped water, tubewells, and bottled water. Townships vary in their access to different sources, but this is not because of income, which appears to be uniform across townships. The water source selection by users depends on intended use and availability, with quality being one of the most important factors in decision-making.

Understanding access to water: multiple sources

In the sample, most (80%) households and businesses obtain water from two or more sources. These include piped water

Table 1 | Sampling plan and survey comparison

	Sampling plan			Actual survey		
Township	**HHs**	**MSMEs**	**Total**	**HHs**	**MSMEs**	**Total**
Amarapura	193 (80%)	48 (20%)	241	193 (80%)	48 (20%)	241
Aungmyethazan	215 (80%)	54 (20%)	269	208 (78%)	60 (22%)	268
Chanayethazan	159 (80%)	40 (20%)	199	160 (80%)	40 (20%)	200
Chanmyathazi	230 (80%)	57 (20%)	287	236 (81%)	57 (19%)	293
Mahaaungmye	194 (80%)	49 (20%)	243	190 (79%)	49 (21%)	239
Pyigyitagon	193 (80%)	48 (20%)	241	191 (80%)	4 (20%)	239
Total	1,184 (80%)	296 (20%)	1,480	1,178 (80%)	302 (20%)	1,480

HH, households; MSMEs, micro, small, and medium enterprises.

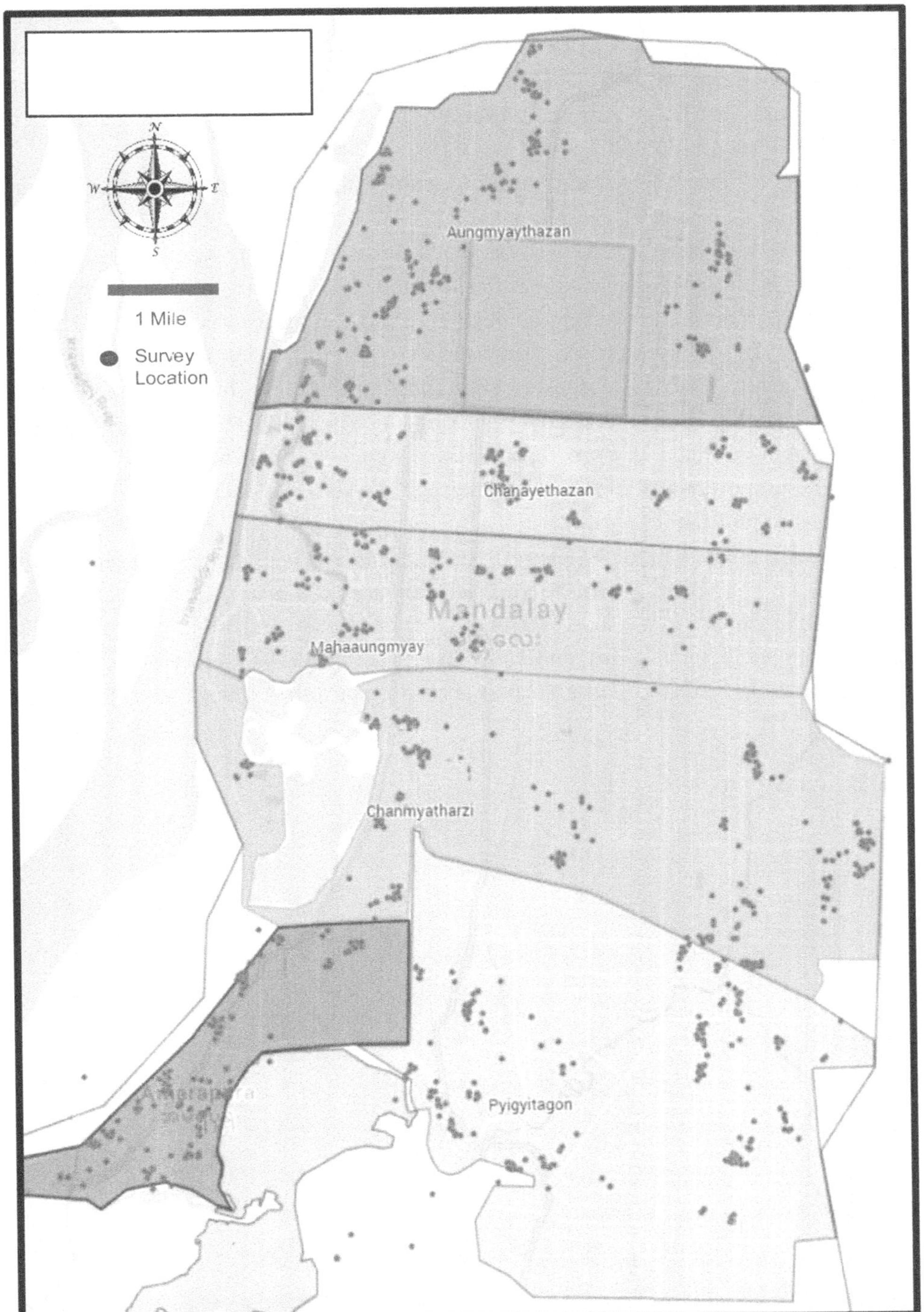

Figure 1 | Map of the survey sample. Surveys, the location of which is represented by a red dot on the map, included households and MSMEs from six townships in Mandalay. Please refer to the online version of this paper to see this figure in colour: http://dx.doi.org/10.2166/washdev.2020.076.

from MCDC, tubewells, or hand-dug wells, commercial bottled water, and other sources, including open sources. The surveyed townships all displayed similar reliance on more than one source. Moreover, water sources are used in similar proportions between households and businesses. Overall, the most common sources of water are water bottles (71%) and tubewells (68%), while only 35% of respondents reported obtaining water through MCDC piped water systems (Figure 2(a)). MSMEs displayed little variation from households in their water sources (Figure 2(b)). A small percentage of respondents obtain their water from other sources. Almost 10% get water from public taps. Of those who collect water from public taps, 92% perform the collection daily and spend an average of 12 min per trip.

Access to water sources differed greatly across townships (Figure 3). In Amarapura, 96% of households used water from tubewells, and none reported having a piped water connection. There was an inversely proportional relationship between use of MCDC piped water and tubewells for households and the similar pattern holds for MSMEs.

Despite the variation across townships in access to different water sources, households displayed marked homogeneity in income distribution (Figure 4(a)). Most households fell between K300,000 and K500,000 a month (USD $210–$350). Nevertheless, water sources used by households do not appear contingent on the level of income (Figure 4(b)).

Understanding water usage: patterns and perceptions

The survey aimed to assess water usage and decision-making patterns. Almost all primary sources were either tubewells (48.6%) or MCDC piped water (44.7%). Households reportedly used their primary source for multiple purposes: showering (97.4%), washing dishes (98.4%), and some cooking (65.5%). Only about a third (27.2%) used their primary source – piped water or tubewells – for drinking despite 42.4% of households saying it was safe to do so.

Figure 5 summarizes the household water usage patterns across all primary, secondary, and tertiary water sources. Households clearly rely on bottled water for drinking, and MCDC water or tubewells for other water-intensive tasks, such as showering or dish washing.

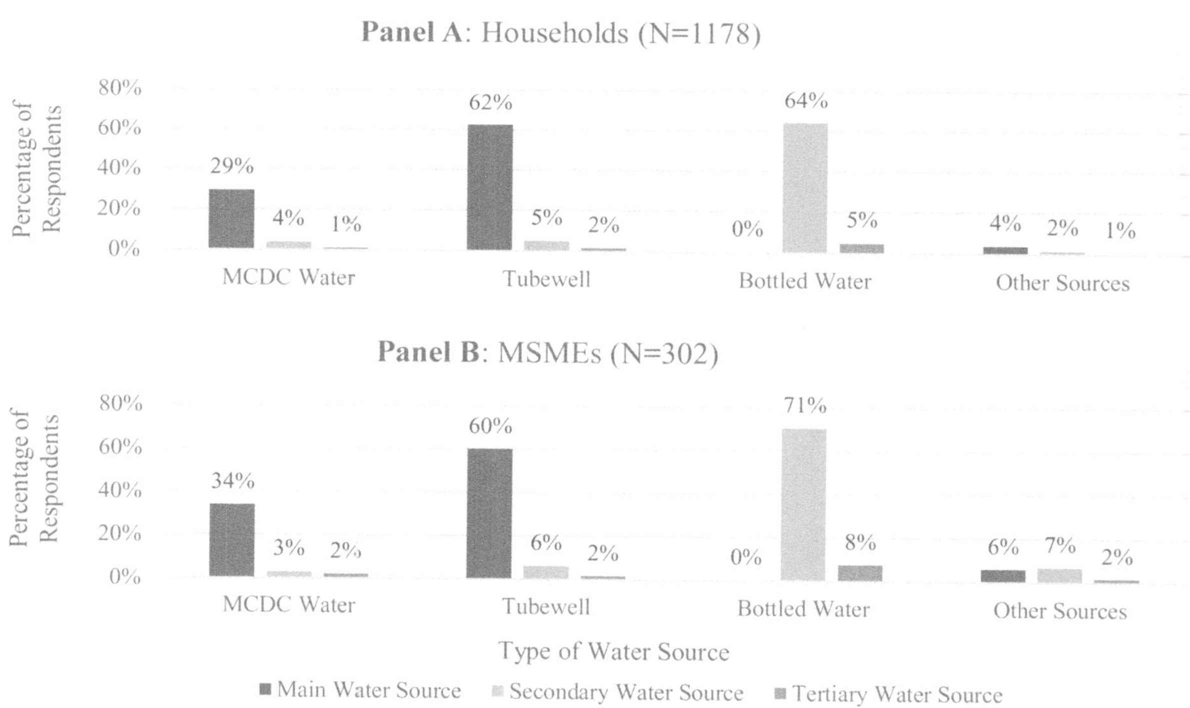

Figure 2 | Primary, secondary, and tertiary water sources used in Mandalay City. Other sources include public tap water, water sold by any other means, protected hand-dug well, protected spring/pond/rainwater/unprotected hand-dug well, unprotected spring/pond/rainwater, and river/stream/lake/cam.

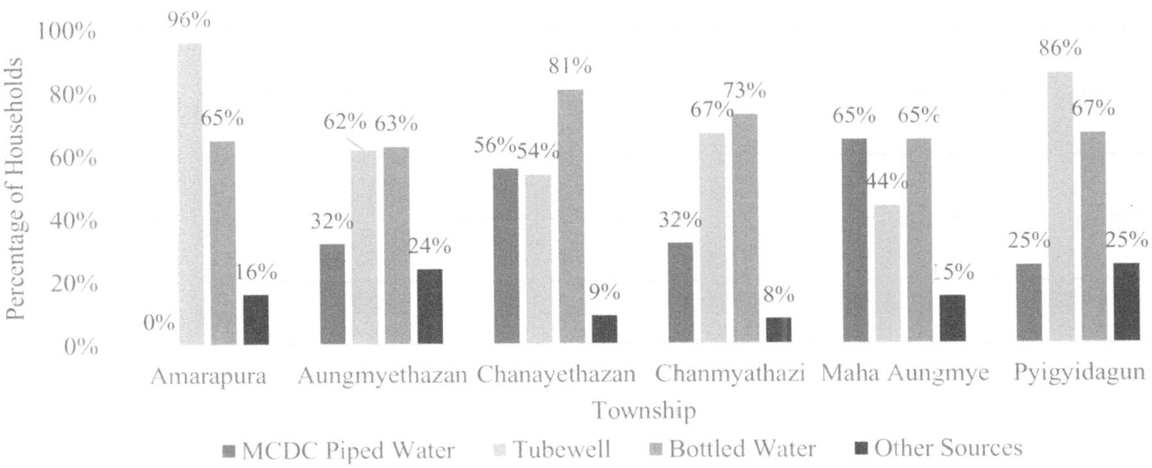

Figure 3 | Household primary, secondary, and tertiary water sources across townships (*N* = 1,178). Other sources include public tap water, water sold by any other means, protected hand-dug well, protected spring/pond/rainwater/unprotected hand-dug well, unprotected spring/pond/rainwater, and river/stream/lake/dam.

Respondents across townships highly valued water quality for consumption. Almost 80% of all households had a second water source, used primarily for drinking (91.8%). This usage was heavily supplied by commercial bottled water – more than 80%, compared to 11% by MCDC piped water and 6% by tubewells. Households used commercial bottled water regardless of whether they had access to the city's piped water or tubewells.

Mandalay citizens emphasized the quality and over half (54%) of respondents stated *quality of water* (smell, taste, etc.) as 'very important'. Only a third responded that *convenience of access* (36%), *reliability of access* (31.4%) or, interestingly, the *price of water* (20.3%), were 'very important'. In general, price and quality seldom factored into why people did not use MCDC piped water. Instead, the decision stemmed from whether MCDC service was available in their area or if they had already invested in a substitute. Based on the survey results, tubewells and MCDC piped water appear to be substitute sources dependent on access, and commercial bottled water a necessary supplement for drinking due to water quality in the city.

Understanding household billing

Among the sampled households, 95% reported a typical monthly income of less than 800,000 MMK (560 USD), and the mean income was K300,000–K500,000 ($210–$350). Only 40% of households reportedly paid for their primary source of water. On average, those who used MCDC piped water paid about 7,668.4 MMK ($5.37) for their primary water source per pay period, and most paid every 2 months (67%) or monthly (19%).

In addition to their primary source, many households (62%) also had commercial bottled water delivered to their homes (12% of households collected it themselves). Most households reported paying between 200 and 2000 MMK ($0.14–1.40) per delivery, with varying frequency of deliveries. Assuming weekly deliveries, or four times a month, a household may pay between 800 and 8,000 MMK ($0.56–5.60) a month on bottled water. Assuming a daily delivery on the other hand, or 30 times a month, bottled water could cost a household between K6,000 and K60,000 ($4.20–42.00 USD). With a monthly income of K400,000 ($280), commercial bottled water delivery could therefore comprise between 0.2 and 15% of monthly expenditure.

According to MCDC officials, the connection fee to the piped water network is about 150,000 MMK (around 105 USD). The survey revealed this connection fee, not the tariff rate, deterred users from connecting to the network. Only 11% of unconnected households and MSMEs reported that cost deterred them from using MCDC piped water. A third of those currently not using MCDC water said they did not need it. Another 71% reported that MCDC water was not available in their area; thus, one can assume that they paid little to nothing for their primary source (only 40% pay for their primary source). Households connected

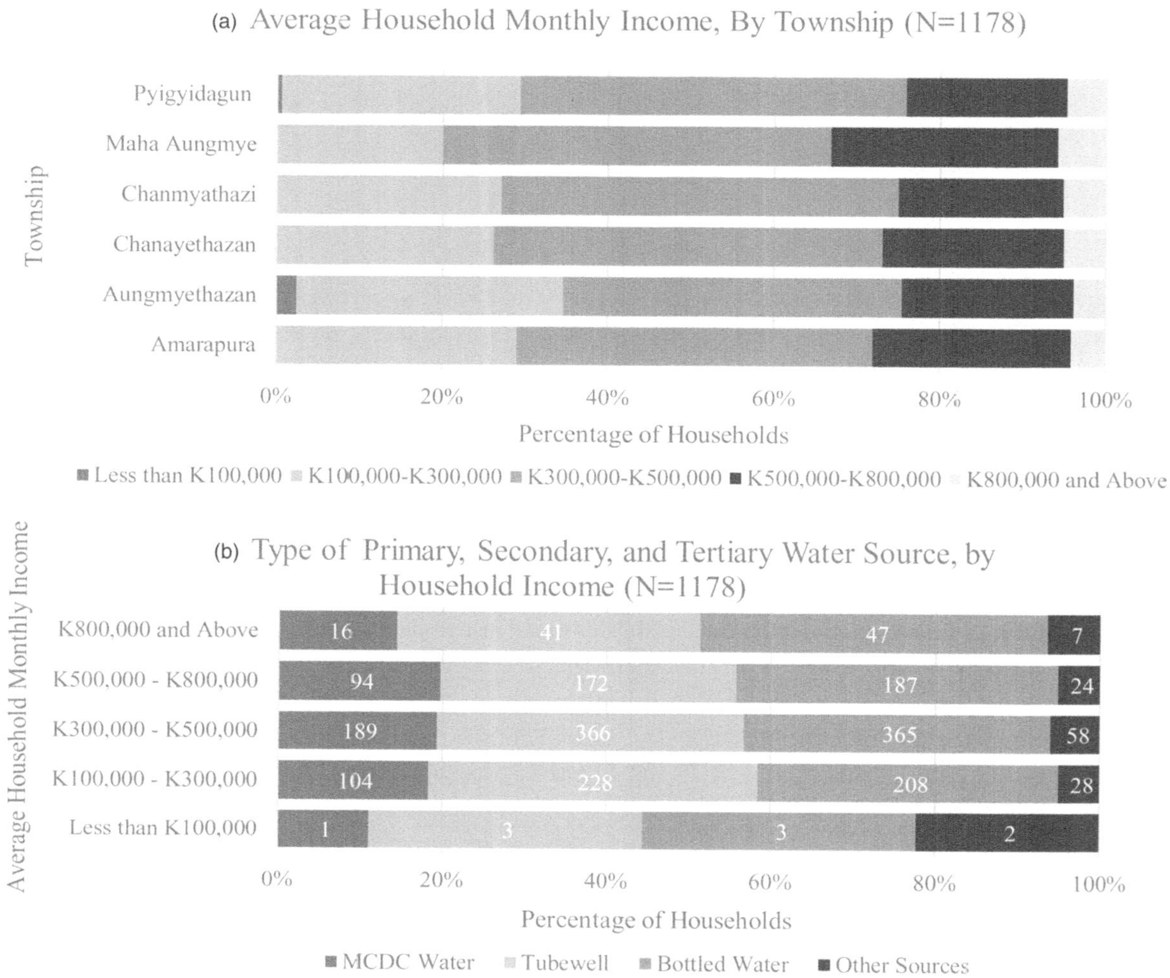

Figure 4 | Household income and water source. Other sources include public tap water, water sold by any other means, protected hand-dug well, protected spring/pond/rainwater/unprotected hand-dug well, unprotected spring/pond/rainwater, and river/stream/lake/dam. *Note:* In (b), the small number of households earning an average monthly income of less than K100,000 makes drawing conclusions about their water source patterns difficult. It should not necessarily be interpreted that this income bracket relies on other sources more than different income groups.

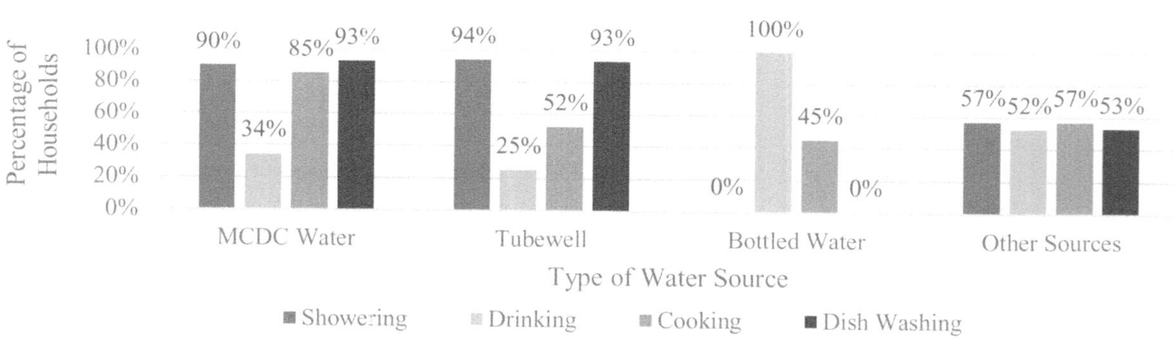

Figure 5 | Water usage, by type of water source ($N = 1,178$). Other sources include public tap water, water sold by any other means, protected hand-dug well, protected spring/pond/rainwater/unprotected hand-dug well, unprotected spring/pond/rainwater, river/stream/lake/dam.

to the MCDC network report that they were billed on a per unit basis, though some were unsure. Of all respondents, the survey found about 36% had a meter and 92% were reportedly functioning. According to MCDC officials, however, about 50% of meters are not read because they are unreadable, not working or broken. MCDC has a team of 55 m readers. If a meter is broken or unreadable, MCDC billed the household based on the average consumption over the last 3 months.

DISCUSSION

This research filled a critical data gap by producing reliable primary data on water supply and usage in Mandalay City. Such data are critical input for policymakers seeking to revise tariffs to fund service improvements. From the primary research, we conclude that users rely on either tubewells or MCDC piped water for uses other than drinking, and these sources are viewed as substitutable. Commercially bottled water is used for drinking (and sometimes cooking) irrespective of income and access to MCDC pipes. Even though some households responded that water quality of their primary source is acceptable, neither MCDC piped nor tubewell water is regarded as safe enough to drink across all income categories. It is difficult to say whether piped water is safe to drink as there is a lack of testing and reliable data on water quality. In the context of the Sustainable Development Goals, over 90% of Mandalay's population has access to improved water. However, because this water may not be 'free from contamination', it does not meet the standard of safely managed drinking water (World Bank 2017). According to one report, 'Piped water supply systems in the main cities of Mandalay and Yangon include untreated surface water from open reservoirs, thereby falling between the Joint Monitoring Program's definitions of improved water and unimproved water' (ADB 2017a). The latest investment in water supply does include plans for the purchase and installation of chlorination equipment (Based on JICA (no date) project preparation documents available online https://openjicareport.jica.go.jp/pdf/12231130_02.pdf and Myanmar Times (2018) newspaper report https://www.mmtimes.com/news/italian-firm-purify-water-mandalay.html (accessed May 29, 2020).). The high reliance of businesses and households on tubewells is also an important finding, especially since tubewells are dug without permission and may exacerbate groundwater depletion and scarcity. In Amarapura, which is urbanizing rapidly, 96% of the sample surveyed relied almost completely on tubewells. A recent study on groundwater shows that there is a high rate of contamination in wells in Amarapura (Grzybowski *et al.* 2019). Thus, as the government expands its public water supply, it will have to consider both its own use of wells and more strictly regulate drilling and extraction by households and MSME.

IDIs conducted with key MCDC officials revealed that billing data are not available for all households and all periods. However, tariffs were revised in 2015 from 55 to 85 MMK per unit (+54.5%). In 2017, the tariff was raised again to 200 MMK (+135.3%). Both tariff increases are expected to have raised MCDC revenues, but it is difficult to establish with certainty the impact that these tariff increases had on household consumption since metering is uneven, and billing records are not easy to find or comparable across time (all billing is done on paper and tariffs are collected in person). It is possible that marginal returns to further sharp tariff hikes could be very low and even negative (Balac *et al.* 2019). Furthermore, given that so many users pay a flat rate per unit (approximated either by household size or historical use), an increase in tariff would be regressive and hurt the poorest MCDC consumers. One other way to address cost recovery could be through reducing non-revenue water, which administrators estimated to be as high as 55%. About 70% of Mandalay's NRW may stem from physical losses and the rest from commercial loss – specifically broken and unreadable meters or faulty meter readings (In 2019, MCDC began the process of testing automatic meter readers, but it is unclear if large scale upgrading of meters has been completed at this time.). Schools, all government buildings, and monasteries are not metered and receive water at no charge. MCDC will need to consider these challenges as it prepares for a growing population and service expansion.

Finally, it is important to remember that piped water coverage is still quite limited; slightly more than a third (35%) of Mandalay City residents use MCDC water as a primary source of water, and only 7% as a second or third option. If MCDC must rely solely on user charges to repay its loans to development partners and to cover operational costs, it will need to significantly expand coverage in addition to raising tariffs.

CONCLUSION

Mandalay City faces a future of population growth and climate challenges that will stress municipal water provision. Households and MSMEs in six townships currently rely on multiple water sources and largely use only bottled water for drinking. As the city administration considers tariff readjustments to finance supply expansion, it must take into account the widespread availability and use of tubewell water as a substitute to piped water, and the supplemental commercial bottled water used for drinking. While households lacking access to tubewells will have to pay the higher tariffs, those with tubewells may substitute MCDC water more often with increasingly scarce, and perhaps contaminated, tubewell water. In the short to medium term, MCDC may need to focus on raising revenues by improving the frequency and reliability of metering and better detection of leaks. Regulating groundwater abstraction may also be viewed as an immediate priority as Mandalay lies in the most arid region of Myanmar and is already facing climate change-related water scarcity.

ACKNOWLEDGEMENTS

This study was part of a recent International Growth Centre (IGC) project conducted for the Mandalay City Development Committee to inform policy decisions on service expansion. It was funded by IGC and carried out under the management of Athena Infonomics by a team from the Johns Hopkins University School of Advanced International Studies, Athena Infonomics and IGC. The authors are grateful to Nilar Win, Nan Sandi, Astrid Haas, Ian Porter (IGC), and Ankit Chatri, Akshay Natteri, Kun Zhang (Athena Infonomics) for their support.

DATA AVAILABILITY STATEMENT

All relevant data are included in the paper or its Supplementary Information.

REFERENCES

Arbués, F., García-Valiñas, M. Á. & Martínez-Espiñeira, R. 2003 Estimation of residential water demand: a state-of-the-art review. *The Journal of Socio-Economics* **32** (1), 81–102.

Asia Development Bank (ADB) 2016 *Loan Agreement (Special Operations) for Loan 3316-MYA: Mandalay Urban Services Improvement Project.*

Asia Development Bank (ADB) 2017a *Myanmar: Urban Development and Water Sector Assessment, Strategy, and Road map.* Metro Manila, Philippines.

Asia Development Bank (ADB) 2017b *Preparing the 2nd MUSIP – Technical Assistance Report.* Metro Manila, Philippines.

Balac, M., Chhatri, A., Nagpal, T., Natteri, A. & Rawlings, H. 2019 *Understanding Demand and Funding for Piped-Water Supply in Mandalay City. Final Report.* International Growth Center C-53448-MYA-1.

Basani, M., Isham, J. & Reilly, B. 2008 The determinants of water connection and water consumption: empirical evidence from a Cambodian household survey. *World Development* **36** (5), 953–968.

Basu, M., Hoshino, S., Hashimoto, S. & Dasgupta, R. 2017 Determinants of water consumption: a cross-sectional household study in drought-prone rural India. *International Journal of Disaster Risk Reduction* **24**, 373–382.

Casey, J. F., Kahn, J. R. & Rivas, A. 2006 Willingness to pay for improved water Service in Manaus, Amazonas, Brazil. *Ecological Economics* **58** (2), 365–372.

Central Statistical Organization (CSO), UNDP and World Bank 2018 *Myanmar Living Conditions Survey 2017: Key Indicators Report.* Nay Pyi Taw and Yangon, Myanmar: Ministry of Planning and Finance, UNDP and the World Bank Group.

Department of Population (DOP), Ministry of Immigration and Population 2015 *The 2014 Myanmar Population and Housing Census: Myanmar Region.* Republic of the Union of Myanmar, Nay Pyi Taw. (3)1.

Eckstein, D., Kunzel, V., Winges, M. & Schafer, L. 2019 *Global Climate Risk Index 2019.* Germanwatch e.V, Bonn, Germany.

Ford, W., Owen, J. & Pillai, S. 2019 *Insight Into Urban Well-Being in Myanmar: The 2018 City Life Survey.* The Asia Foundation, San Francisco, CA.

Gross, E. & Elshiewy, O. 2019 Choice and quantity demand for improved and unimproved public water sources in rural areas: Evidence from Benin. *Journal of Rural Studies* **69**, 186–194.

Grzybowski, M., Lenczewski, M. E. & Oo, Y. Y. 2019 Water quality and physical hydrogeology of The Amarapura township, Mandalay, Myanmar. *Hydrogeology Journal* **27**, 1497–1513.

International Monetary Fund (IMF) Asia and Pacific Dept 2019 *Myanmar: 2018 Article IV Consultation-Press Release; Staff Report; and Statement by the Executive Director for Myanmar.* International Monetary Fund, Washington, DC.

Lavrakas, P. J. 2008 *Encyclopedia of Survey Research Methods,* Vols. 1–0. Sage Publications, Inc., Thousand Oaks, CA.

Lombana Cordoba, C., Thibert, M. D., Danilenko, A. V., Andres, L. A., Borja-Vega, C. & Jospeh, G. 2019 *Doing More with Less: Smarter Subsidies for Water Supply and Sanitation*. World Bank, Washington, DC, p. xiii.

Martínez-Santos, P. 2017 Determinants for water consumption from improved sources in rural villages of Southern Mali. *Applied Geography* **85**, 113–125.

Marzano, R., Rougé, C., Garrone, P., Grilli, L., Harou, J. J. & Pulido-Velazquez, M. 2018 Determinants of the price response to residential water tariffs: meta-analysis and beyond. *Environmental Modelling & Software* **101**, 236–248.

Mitlin, D. & Walnycki, A. 2019 Informality as experimentation: water utilities' strategies for cost recovery and their consequences for universal access. *The Journal of Development Studies* **9** (4), 259–277.

Nagpal, T., Eldridge, M. & Malik, A. A. 2019 Global water access fund: a new idea to bridge operations and maintenance shortfalls for the poorest water utilities. *Journal of Water, Sanitation and Hygiene for Development* **9** (4), 774–779.

Nauges, C. & Whittington, D. 2010 Estimation of water demand in developing countries: An overview. *World Bank Research Observer* **25** (2), 263–294.

North, J. H. & Griffin, C. C. 1993 Water source as a housing characteristic: hedonic property valuation and willingness to pay for water. *Water Resources Research* **29** (7), 1923–1929.

Orgill, J., Shaheed, A., Brown, J. & Jeuland, M. 2013 Water quality perceptions and willingness to pay for clean water in peri-urban Cambodian communities. *The Journal of Water and Health* **11** (3), 489–506.

Pattanayak Subhrendu, K., Yang, J.-C., Whittington, D. & Bal Kumar, K. C. 2005 Coping with unreliable public water supplies: averting expenditures by households in Kathmandu, Nepal. *Water Resources Research* **41** (2), W02012.

Pink, R. M. 2016 *Water Rights in Southeast Asia and India*. Palgrave Macmillan, New York.

Rab, H. N., Zachau, U., Seck, A., Verghis, M. A. & Mohib, S. S. A. 2016 *All Aboard!: Policies for Shared Prosperity in Myanmar (English)*. World Bank Group, Washington, DC.

Silva Pinto, F. & Cunha Marques, R. 2015 Tariff structures for water and sanitation urban households: a primer. *Water Policy* **17** (6), 1108–1126.

Skinner, C. J. 2016 Probability proportional to size (PPS) sampling. In: *Wiley StatsRef: Statistics Reference Online* (N. Balakrishnan, T. Colton, B. Everitt, W. Piegorsch, F. Ruggeri & J. L. Teugels, eds). John Wiley & Sons, Ltd, pp. 1–5.

State Peace and Development Council 2002 *The City of Mandalay Development Law (The State Peace and Development Council Law No. 8/2002)*. Rangoon, Myanmar.

Tham, T. N. & Department of Meteorology and Hydrology, Ministry of Transportation 2010 *Some Observed Climate Change Impacts in Myanmar*. Republic of the Union of Myanmar, Nay Pyi Taw, Myanmar.

United Nations Children s Fund (UNICEF) 2017 *Thirsting for A Future: Water and Children in A Changing Climate*. UNICEF, New York, NY.

United Nations, Department of Economic and Social Affairs (UNDESA) 2018 *World Urbanization Prospects: The 2018 Revision*. Available from: https://www.un.org/development/desa/publications/2018-revision-of-world-urbanization prospects.html.

Van Den Berg, C. & Danilenko, A. V. 2017 *Performance of Water Utilities in Africa (English). Water Global Practice*. World Bank Group, Washington, DC.

Van Houtven, G. L., Pattanayak, S. K., Usmani, F. & Yang, J. 2017 What are households willing to pay for improved water access? results from a meta-analysis. *Ecological Economics* **136**, 126–135.

Wagner, J., Cook, J. & Kimuyu, P. 2019 Household demand for water in Rural Kenya. *Environmental and Resource Economics*, **74**(4), 1563–1584. Springer; European Association of Environmental and Resource Economists.

Whittington, D. & Nauges, C. 2010 Estimation of water demand in developing countries: an overview. *World Bank Research Observer* **25**, 263–294.

World Bank 2017 *Myanmar Living Conditions Survey 2017. Report 03: Poverty Report*. World Bank and UNDP. Available from: http://documents.worldbank.org/curated/en/921021561058201854/pdf/Myanmar-Living-Condition-Survey-2017-Report-3-Poverty-Report.pdf (accessed 6 January 2020).

Worthington, A. C. & Hoffmann, M. 2006 *A State of the Art Review of Residential Water Demand Modelling, University of Wollongong, School of Accounting and Finance Working Paper Series No. 06/27*.

WWAP (UNESCO World Water Assessment Programme) 2019 *The United Nations World Water Development Report 2019: Leaving No One Behind*, Vol. 87. UNESCO, Paris.

First received 30 March 2020; accepted in revised form 11 July 2020. Available online 6 August 2020

Research Paper

Unpacking piped water consumption subsidies: Who benefits? New evidence from 10 countries

Laura Abramovsky, Luis Andrés, George Joseph, Juan Pablo Rud, Germán Sember and Michael Thibert

ABSTRACT

This paper provides new evidence on the recent performance of piped water consumption subsidies in terms of pro-poor targeting for 10 low- and middle-income countries around the world. Our results suggest that in these countries, existing tariff structures fall well short of recovering the costs of service provision, and that, moreover, the resulting subsidies largely fail to achieve the goal of improving the accessibility and affordability of piped water among the poor. Instead, the majority of subsidies in all 10 countries are captured by the richest households. On average, across the 10 low- and middle-income countries examined, 56% of subsidies end up in the pockets of the richest 20%, but only 6% of subsidies find their way to the poorest 20%. This is predominantly due to the most vulnerable segments of the population facing challenges in access and connection to piped water services. Shortcomings in the design of the subsidy, conditional on poor households being connected, exist but are less important.

Key words | consumption of piped water, developing country, distributional incidence of subsidies

Laura Abramovsky (corresponding author)
Institute for Fiscal Studies (IFS) and MacroConsulting,
7 Ridgmount Street, London WC1E 7AE,
UK
E-mail: *labramovsky@ifs.org.uk*

Luis Andrés
George Joseph
Michael Thibert
World Bank (WB),
WB 1818 H Street, NW, Washington, DC 20433,
USA

Juan Pablo Rud
Royal Holloway, University of London and IFS,
Egham Hill, Egham TW20 0EX,
UK

Germán Sember
MacroConsulting,
Lavalle 190, Piso 5° Oficina L, Buenos Aires,
Argentina

HIGHLIGHTS

- This study provides new evidence on the recent performance of piped water consumption subsidies in terms of pro-poor targeting for 10 low- and middle-income countries around the world: Ethiopia, Mali, Niger, Nigeria, Uganda, El Salvador, Jamaica, Panama, Bangladesh, and Vietnam.
- The findings suggest that piped water consumption subsidies in all 10 countries tend to be regressive and therefore do not adequately target the poor.
- The majority of subsidies in all 10 countries are captured by the richest households. On average, across the 10 low- and middle-income countries examined, 56% of subsidies end up in the pockets of the richest 20%, but only 6% of subsidies find their way to the poorest 20%.
- This is predominantly due to the poorest 40% facing challenges in access and connection to piped water services.

doi: 10.2166/washdev.2020.073

GRAPHICAL ABSTRACT

Distributional incidence of subsidies (by decile) for five African countries

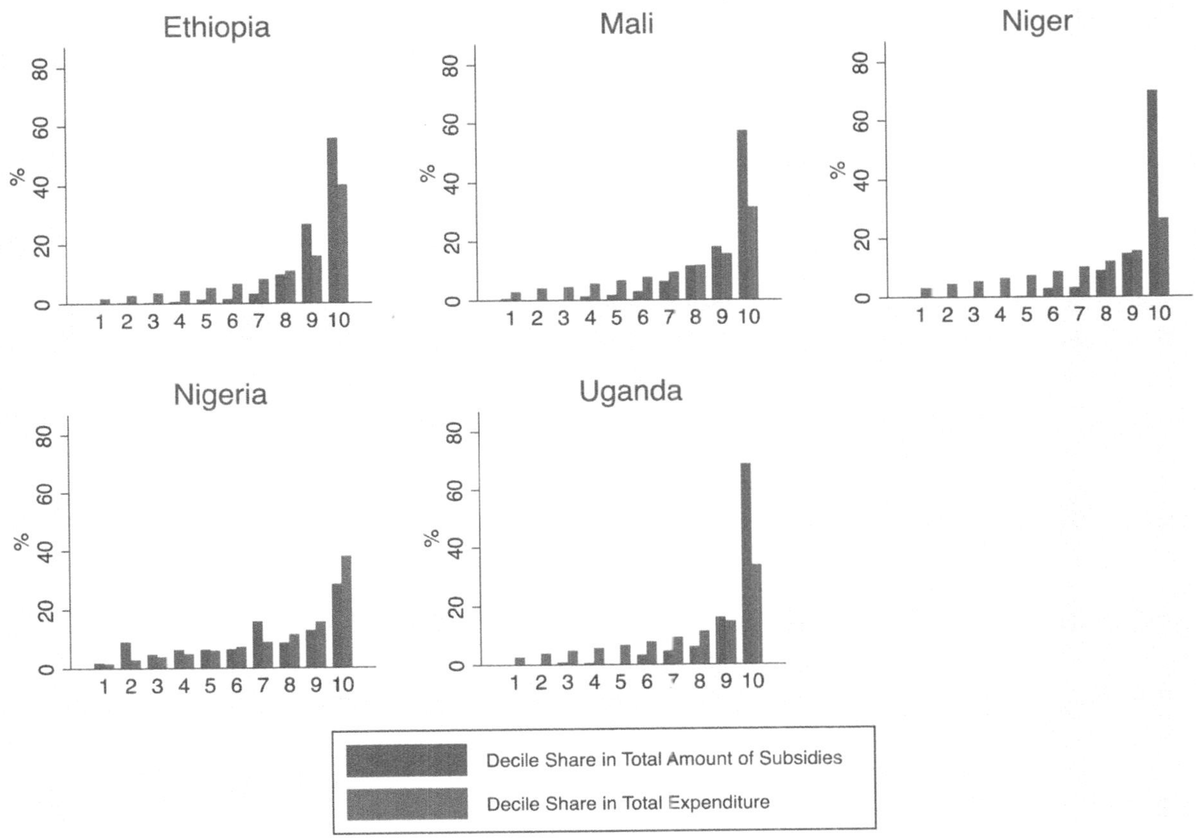

INTRODUCTION

Universal access to water and sanitation are stated international sustainable development goals (SDGs). The United Nations adopted these SDGs in 2015 with the aim to 'achieve universal and equitable access to safe and affordable drinking water' and to 'achieve access to adequate and equitable sanitation and hygiene for all' by 2030. However, many countries around the world are still struggling to achieve this goal. In 2015, 29% of the world population lacked access to safely managed water and 61% lacked safely managed sanitation (WHO/UNICEF 2017). The cost of meeting these gaps is estimated to be around $100 billion a year (Hutton & Varughese 2016). Well-designed subsidies stand out as a key instrument to achieve these SDGs, alongside other policies such as investment, technological innovations or better governance and planning. In the context of high levels of poverty, where markets alone do not result in the desired levels of service provision and consumption, targeted subsidies can help address affordability and equity issues. Additionally, water and sanitation subsidies are generally advocated because access to and consumption of water and sanitation services are associated with rising productivity and living standards, positive externalities related to public health, and may free up time spent collecting water (WWAP 2016; Hutton & Chase 2017). For example, Hutton & Chase (2017) show that health and non-health associated costs due to poor water and sanitation are estimated to be over 5% of GDP in 6 of the 30 countries included in their study.

A subsidy represents the difference between the supplier's cost of providing the service and the actual price or tariff paid by a user. Consumption subsidies reduce the cost of consuming water and, as such, are only available to existing customers. These can be distinguished from connection subsidies, which are a one-time reduction in connection charges and hence only available to new, previously unconnected, customers. Water consumption subsidies can take different shapes and sizes: untargeted subsidies (general underpricing of water supply that benefits all consumers); implicit subsidies (generated by flat fees for unmetered services, low meter coverage or low revenue collection); or explicit subsidies (e.g., quantity targeting using increasing block tariffs (IBTs) or subsidies using administrative selection, such as means-tested or geographic targeting). The funding for these subsidies comes from one or a combination of the following two sources: government transfers and rate funds (that arise from charging other users more than the cost of the service, also called cross-subsidies). Finally, the transfer mechanism could be implemented in different ways. In a demand-side subsidy, usually the most transparent type of subsidy, the government provides a monetary transfer directly to the user, who uses this transfer towards their payments to the service provider. In a supply-side subsidy, the government transfers the money to the provider. Andres *et al.* (2019) and Komives *et al.* (2005) discuss the different types of WASH subsidies in more detail.

The design and performance of piped water subsidies hinges on the industrial structure and technology of piped water production and delivery and is a contentious area, heavily influenced by political economy considerations. An important challenge for policymakers is to design a subsidy scheme that allows for the recovery of capital and expenditure costs of providing the service (WWAP 2019). Piped water is usually considered a local natural monopoly. The supplier faces large upfront (fixed) costs (resulting in increasing returns to scale) on assets that are long-lived. This feature of water supply, as with other similar public utilities, makes pricing difficult. Allocative efficiency requires prices to be set equal to marginal cost, i.e., the good is consumed up to the point where the benefit consumers obtain from the last unit consumed equals the additional cost incurred by the supplier to produce it. However, pricing at marginal cost in a context of high fixed costs would not allow for full cost recovery (as the marginal cost is lower than the average costs). This feature generates heated debates around how to implement low water tariffs (i.e., close to the marginal cost) while ensuring cost recovery and avoiding financial unviability of utilities. (In the context of lower-income countries, it is recommended that tariffs cover at least the operating expenditures incurred by suppliers, but even this is often not achieved (see, for instance, Baietti & Curiel 2005; Andres *et al.* 2019).) This, in part, is because it is really difficult to estimate the cost of providing each customer with the service, as costs vary across consumers, depending on geographic location, topography, distance to the source, etc. This situation translates into a high degree of discretion on how to allocate provision costs.

In practice, this often means subsidization of the service across the board. Most countries, whether rich or poor, end up putting in place subsidies schemes funded by the government or designing tariffs so that richer consumers cross-subsidize poorer ones (GWI 2004; Komives *et al.* 2005; Andres *et al.* 2019). New estimates by Andres *et al.* (2019) put total subsidies for networked water and sewerage services at around 0.5% of GDP worldwide, and over 1.5% of GDP for non-advanced economies. In developing countries, this subsidization across the board does not fully address the objectives of universal access to clean water and the affordability of the service, and additionally may distort consumption patterns. Furthermore, high subsidization levels have often undermined the financial sustainability of the system, inducing managers to face soft budget constraints and lowering their financial performance. This, in turn, could reduce the ability of service providers to access commercial finance, which could enable utilities' ability to maintain and expand the service in an affordable manner to new, likely poorer customers (see, for instance, Goksu *et al.* 2017).

In this paper, we provide new evidence on the distributional incidence and pro-poor targeting of subsidies for piped water consumption for 10 low- and middle-income countries around the world: Ethiopia, Mali, Niger, Nigeria, Uganda, El Salvador, Jamaica, Panama, Bangladesh and Vietnam. We also document the gap between estimates of actual paid tariffs and tariffs that would be needed to recover the full cost of providing the service (also called cost-recovery or cost-reflective tariffs). All these countries capture relevant heterogeneity since they reflect different levels of gross domestic product per capita, connection rates, tariff structures and cost-recovery tariffs. The cases considered include quantity-based tariffs and flat rates. Importantly, in these countries, there is the availability of household surveys with measures of water expenditure and overall consumption expenditures, as well as administrative data on tariffs from the utility and new estimates of country-specific cost-reflective tariffs. These estimates are calculated using an improved methodology and new data presented in Andres *et al.* (2019) and discussed in more detail in the next section.

Our results suggest that, in all the countries analyzed, mean unit prices charged are, on average, lower than the overall cost of producing and distributing piped water, resulting in substantial water consumption subsidies. In addition, subsidies tend to be regressive, with the amount of resources allocated to water consumption subsidies increasing over the expenditure distribution, and richer households in the top deciles usually capturing the lion's share. On average, across the 10 low- and middle-income countries examined, 56% of subsidies end up in the pockets of the richest 20%, but only 6% of subsidies find their way to the poorest 20%.

Our findings also highlight the importance of access to the service in explaining the regressive nature of these subsidies. In particular, (i) poor households live in areas that are not covered by piped water networks (poor households are defined as belonging to the first four deciles of the countrywide national expenditure (or income) per capita distribution in each country), (ii) poor households live in areas with coverage but are not connected to the network and (iii) poor households that are connected to the network appear to be consuming smaller quantities of water than the general population. This implies that there are high errors of inclusion (i.e., households not among the 40% poorest receiving subsidies) and even higher errors of exclusion (households among the 40% poorest not receiving subsidies). This issue is particularly pronounced in the considered African countries, where errors of exclusion fall between 90 and 100%.

Our study contributes to a vast and growing literature on the distributional incidence of fiscal policy in low- and middle-income countries, including a large sub-literature on the distributional incidence of water subsidies. Our findings are consistent with a number of previous studies that use a similar methodology. Key examples of this literature – such as Komives *et al.* (2006) and Angel-Urdinola & Wodon (2007b) – have shown that quantity-based, targeted subsidies in Nicaragua, Cape Verde, the city of Kathmandu (Nepal), Bangalore (India) and Sri Lanka are usually regressive, with a smaller share of benefits accruing to the poor than the general population. These studies also indicate that poor targeting is mostly associated with low rates of access to water networks in poor neighborhoods, as well as low connection rates for poor households in neighborhoods with access. More recently, the World Bank (2017) found that Tunisian households in the bottom quintile of the income distribution receive 11% of water subsidies, while the top quintile receives 27%. Our study provides systematic cross-country evidence, using a novel and more robust estimate of cost-recovery tariffs based on actual country-specific water providers' data.

Lack of access to the service by the poor has been identified as an important driver of regressive subsidies in other sectors. There are numerous studies documenting similar patterns in the electricity sector as documented by Komives *et al.* (2006). Recent studies conducted as part of the Commitment to Equity (CEQ) initiative looking at the distributional incidence of fiscal policy in a range of low- and middle-income countries also confirm these findings using a different methodology (see, for instance, Hounsa *et al.* (2019) for energy subsidies in Mali). CEQ studies, discussed in Inchauste & Lustig (2017), find that lack of access by the poor to tertiary education is one of the reasons why public education expenditure is regressive in Ethiopia, Ghana, Guatemala and Indonesia. Access also plays a role in the distributional incidence of health subsidies. For example, Chen *et al.* (2015) find that Chinese government healthcare subsidies are pro-rich and regressive and that lack of access plays an important role. O'Donnell *et al.* (2008) show the distribution of public healthcare subsidies in Vietnam is pro-rich, although the impact varies by the type of healthcare facility.

Finally, quantity-based consumption subsidies have been found to be regressive, even conditional on having access and being connected to the network. For example, Cardenas & Wittington (2019) use data for connected households in Addis Ababa, Ethiopia and find that water and electricity quantity-based subsidies are large and regressive. They use improved measures of quantities consumed by accessing administrative utilities billing data and matching it to household survey data. In contrast, when subsidies are targeted using administrative-based mechanisms (i.e., geographic targeting or means testing), the evidence suggests that these are likely to be more progressive and pro-poor (Komives *et al.* 2006).

The rest of the paper is structured as follows: Section 'Data and methodology' presents the methodology, the data used and related statistics; Section 'Findings' presents the findings and the last section summarizes the findings and discusses the policy implications.

DATA AND METHODOLOGY

In order to estimate consumption subsidies for piped water at the household level, we need a measure of the cost of delivering a unit (cubic meter, m^3) of piped water, a measure of the unit price actually paid by each household and the quantity consumed. To do this, we combine several data sources spanning different periods within and across countries. We use a methodology that closely follows Komives *et al.* (2005) and Angel-Urdinola & Wodon (2007a), to estimate the distributional and targeting performance of subsidies in each country.

We first describe the multiple data sources used, and then explain the methodology and its implementation. We subsequently provide some descriptive statistics before discussing the limitations of the data and methodology used in this study.

Data sources

Household surveys (source 1)

Household level data comes from the latest available socio-economic (or income and expenditure) household surveys collected in each country. Most surveys are from 2015 or 2016, but some surveys are from earlier periods, the earliest being 2012 for Jamaica. Information about the household surveys used for each country considered in the analysis is provided in Supplementary Appendix Table A1.

These are general socio-economic surveys that cover a range of dimensions of a household's characteristics and income and spending patterns, and hence are not focused specifically on water and sanitation. Nonetheless, they provide useful self-reported information on water access, connection and water expenditure in the last month or last year, depending on the survey. Details about the exact variables used in each survey are also provided in the Supplementary Appendix. Quantities of consumed water and prices or tariffs paid by each household are usually not reported. Consequently, we combine household survey data on water expenditure with administrative data from providers and government programs detailing the tariff structure for water and sewerage services to impute water quantities and tariffs, as below.

This approach has been used in other studies, such as in Komives *et al.* (2006) and Angel-Urdinola & Wodon (2007b). However, it is not without limitations. In particular, water expenditure in the last month (some surveys asked about expenditure over the last 12 months) can provide an imperfect proxy for the expenditure on piped water if (i) households use multiple water sources, including technologies for treating non-piped water (in which case it could overestimate piped water expenditure); (ii) meters are not read, shared (Supplementary Appendix Table A2 shows the proportion of households that use the neighbor's tap as their main source of drinking water for the 4 countries for which the data is available: Mali, Niger, El Salvador and Panama. The proportion is low and varies between 1% for all households in Panama and rural households in Niger, to 10% for urban households in Niger and 7% for urban households in Mali) or do not work; or (iii) the water bill includes pro-rated connection charges and arrears (Whittington *et al.* 2015). Importantly, these surveys also enable the construction of total expenditure (and/or income) variables for each household.

Administrative data on tariff structures (source 2)

Information about tariff structures comes from the International Benchmarking Network for Water and Sanitation

Utilities (IBNET) database. Some countries list most of their water providers in IBNET, while others list only a few. For example, IBNET has data on the tariff structures of all water providers in Jamaica, Mali and Niger, since these have only one piped water provider. Meanwhile, for countries such as Ethiopia, Nigeria, Uganda and Bangladesh – where the provision of services is decentralized and there is a multitude of piped water utilities – IBNET data covers a subgroup of these providers. An additional limitation is that the tariff structure data available to the authors for this study may cover a period that is different than the one covered in the household survey for each country, even though tariffs may not be regularly updated. More information about dates covered for each utility in each country is provided in the Supplementary Appendix.

Data on cost-reflective tariffs (source 3)

Cost-reflective tariffs cover the capital and recurrent costs of providing service, including not only the efficient economic cost but also costs arising from inefficiencies of the service provider. We use estimates of cost-reflective tariffs (defined as CRT) that cover both operating expenditure (OPEX) and capital expenditure (CAPEX), obtained from Andres *et al.* (2019).

Their approach entails answering the following question 'What is the long-run incremental cost of providing water and sanitation services for a given company?'. It is based on a simple, utility-wide, bottom-up model used by regulators in many utility sectors. In the water and sanitation sector, the model firm approach is used in Chile and a few other Latin American countries. The authors choose efficient firm estimates from Chile as a benchmark to determine the capital stock per customer. The Chilean tariff law seeks to induce efficiency through the use of incremental cost of development pricing. Decree – D.F.L. No 70/ 1988 – defines this tariff as the 'value equivalent to a constant per unit price which, when being applied to the incremental forecasted demand, generate revenues to cover incremental operation efficient costs and the required investment from an optimized project of expansion of the firm, such that it should be consistent with a net present value (NPV) of the project equal to zero.' As pointed out by Bitrán & Arellano (2005): 'In Chile, to avoid transferring the cost of inefficiencies to users, the rate setting process emulates competitive conditions by using a fictitious company that would theoretically meet demand over the next five years in the most efficient way.' Using the information on operation and maintenance costs for the period 2010–2015 from the IBNET dataset, they estimate efficient cost-reflective tariffs. In addition, this approach accounts for inefficiencies arising from employees and losses, to obtain a total cost-reflective tariff. Despite its stringent assumptions and simplifications, this comprehensive approach improves on the existing estimates that have been used in the literature. For example, the Price Gap Approach uses a single uniform CRT value for all countries of 1,27 USD/m^3. There are a range of papers by IMF staff that have used this approach to calculate implicit subsidies to public utilities. Recently Kochhar *et al.* (2015) have applied it to networked water, while Ebeke & Ngouana (2015) used it to estimate energy subsidies. Another methodology, known as the Hidden Cost Calculator (Ebinger 2004), uses book values of gross assets instead.

Methodology

Definition of variables of interest

In the following subsections, we describe how we classify households by economic status (i.e., poor or not), identify whether they benefit from water consumption subsidies, and estimate the magnitude of the subsidies they receive. All household level statistics are calculated using sampling weights to correct for the bias inherent in representative household surveys (Deaton 2019). Further information on how each variable is constructed for each country is provided in the Supplementary Appendix.

Figure 1 summarizes how we classify households and Table 1 defines the main variables.

Poor households (P) *and all households* (H). Poor households are defined as belonging to the first four deciles of the countrywide national total expenditure (or income) per capita distribution in each country and are denoted by the subscript p. All households in each country are denoted by the subscript h.

Total expenditure includes expenditure on all goods and services, including the water bill, which is a common

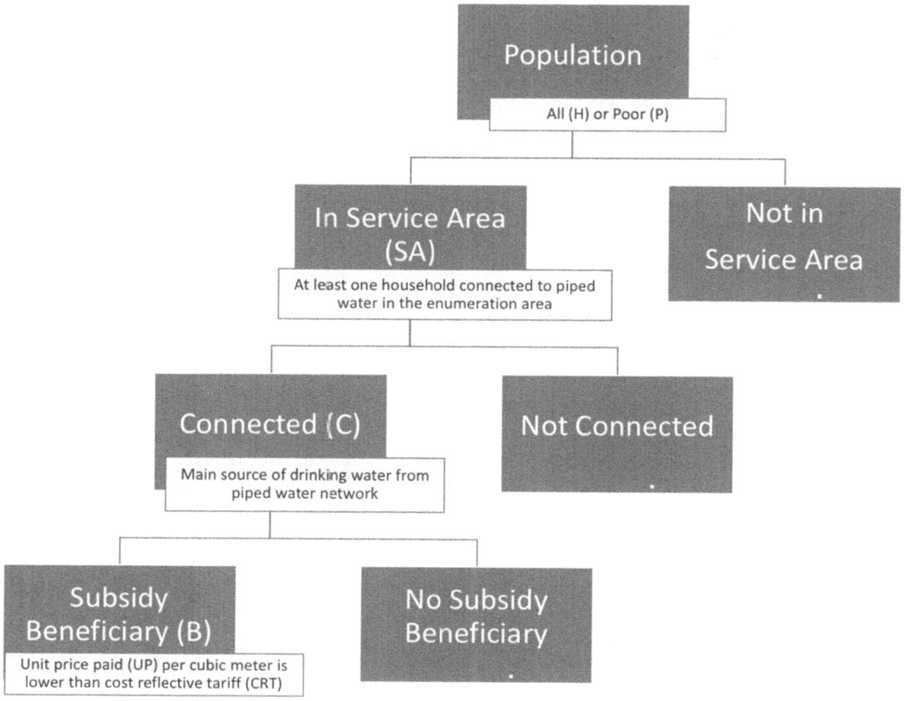

Figure 1 | Classification of households according to piped water access, connection and consumption subsidy beneficiary.

methodology to rank households according to their resources in developing countries (see, for example, Abramovsky & Phillips (2015)). Annual expenditure figures are converted to equivalent monthly figures. It is important to note that although we conduct our analysis at the household level, we construct deciles of expenditure (or income) on the basis of a household's per capita expenditure, not overall household expenditure. Because we conducted our analysis of expenditure at the household level – and average household size may vary across deciles, particularly in Africa – average total expenditure at the household level may not increase as expected when moving across the household per capita expenditure distribution.

Service area (SA). Following Angel-Urdinola & Wodon (2007a), households are considered to have potential access to a network if they are located in a service area (SA_h). (The variable *service area* (SA) is equivalent to the variable *access* (*A*) in Komives *et al.* (2005) and Angel-Urdinola & Wodon (2007a).) In this paper, we prefer to use *service area* because it better reflects the variable we are measuring. A household's service area takes the value of 1 if at least one household in their neighborhood self-reported having a water connection in the household survey. We assume that households in an SA have the option of connecting to the piped water network present in their neighborhood. The neighborhood is proxied by the enumeration area where the household is located, according to the household surveys. However, this assumption may not always be correct. If a neighborhood covers a large geographical area, or if the presence of an adjacent water main varies by household, the survey figures may overestimate the actual number of households with potential access to the piped water network.

Connected (C). We set this variable to 1 if a household is connected to the water network, and to zero otherwise. This is determined by the self-reported main source of drinking water in the household surveys.

Quantity of water consumed among those connected (Q). We construct $Q_{h|C}$, the monthly water quantity consumed (measured in m^3) by imputing water consumption volumes from self-reported household monthly water

Table 1 | Main variables of interest

Variable	Definition	Sources
P (Poor)	= 1 if a household belongs to the first four deciles of the countrywide national total expenditure (or income) per capita distribution in each country. Subscript p is to indicate it is a poor household. Letter P is the total number of poor households	1
H (All)	= 1 for all households. Subscript h indicates any individual household. Letter H is the total number of all households	1
Service area (SA_h)	= 1 if a household h is located in a neighborhood with at least one other household with a water connection as self-reported in the household survey. Neighborhood is proxied by the enumeration area	1
Connected C_h	= 1 if a household h is connected to the water network, and zero otherwise. This is determined by the self-reported main source of drinking water	1
$E_{h\|C}$	Monthly water expenditure for household h, conditional on being connected	1
Quantity of water consumed ($Q_{h\|C}$)	Estimated monthly water consumed (in m^3) conditional on being connected	1 and 2
$\mathrm{UP}_{h\|C}$	Unit price for a cubic meter of water for a household h: $E_{h\|C}/Q_h$	1 and 2
B_h	= 1 if a household h receives a subsidy: $\mathrm{UP}_{h\|C} < \mathrm{CRT}$	1, 2 and 3
B_P	Total number of poor households receiving a subsidy (poor beneficiaries)	1, 2 and 3
B_H	Total number of households receiving a subsidy (all beneficiaries)	1, 2 and 3
R_h	The rate of subsidization for a household h: $1 - \mathrm{UP}_h/\mathrm{CRT}$	1, 2 and 3
S_h	Amount of subsidy received by a household h: $Q_h\|_B \times (\mathrm{CRT} - \mathrm{UP}_h)$	1, 2 and 3
S_P	Total amount of subsidies accrued to the poor	1, 2 and 3
S_H	Total amount of subsidies accrued to all households	1, 2 and 3
Variables to construct omega		
SA_P	% of poor households that are located in a service area	1, 2 and 3
SA_H	% of all households that are located in a service area	1, 2 and 3
$C_{\mathrm{P\|SA}}$	% of poor households that are connected conditional on being located in a service area	1, 2 and 3
$C_{\mathrm{H\|SA}}$	% of all households that are connected conditional on being located in a service area	1, 2 and 3
$B_{\mathrm{P}\|C}$	% of poor households with a subsidy, conditional on being connected	1, 2 and 3
$B_{\mathrm{H}\|C}$	% of all households with a subsidy, conditional on being connected	1, 2 and 3
$R_{\mathrm{P}\|B}$	Average rate of subsidization for poor households, conditional on receiving a subsidy	1, 2 and 3
$R_{\mathrm{H}\|B}$	Average rate of subsidization for all households, conditional on receiving a subsidy	1, 2 and 3
$Q_{\mathrm{P}\|B}$	Average quantities consumed by poor households, conditional on receiving a subsidy	1, 2 and 3
$Q_{\mathrm{H}\|B}$	Average quantities consumed by all households, conditional on receiving a subsidy	1, 2 and 3
Ω	$\Omega = \frac{S_\mathrm{P}/P}{S_\mathrm{H}/H} = \frac{\mathrm{SA_P}}{\mathrm{SA_H}}\frac{C_{\mathrm{P\|SA}}}{C_{\mathrm{H\|SA}}}\frac{B_{\mathrm{P}\|C}}{B_{\mathrm{H}\|C}}\frac{R_{\mathrm{P}\|T}}{R_{\mathrm{H}\|T}}\frac{Q_{\mathrm{P}\|T}}{Q_{\mathrm{H}\|T}}$	1, 2 and 3

expenditure, $E_{h|C}$, conditional on being connected to the network. The conversion is undertaken using the corresponding tariff structure from the IBNET data. (Information on how this match is done for each country is provided in the Supplementary Appendix.)

In the first round of calculations, we impute quantities for households for which (i) the total monthly bill for water consumption depends on quantities consumed, (ii) the value of self-reported expenditure on water is greater than zero and (iii) we can map the household to a specific provider for which we have tariff structure information. For households that pay a fixed rate per cubic meter, the monthly quantity of water consumed equals the total expenditure on water divided by the rate. For households facing unit prices that vary by quantity consumed (such as increasing block or volume-differentiated tariffs), each block within the tariff structure is assigned a maximum expenditure level and associated quantity of water consumed. Then, quantities are assigned to the household sample by matching self-reported household water expenditure with the corresponding level of the tariff structure data.

For the remainder of the households for which (i) the bill is a flat rate unrelated to water quantities consumed or (ii) self-reported expenditure on water equals zero, we assume consumption quantities at the median of those households in the same country for which we could impute quantities. In many countries, and particularly Ethiopia and Nigeria, many households appear to be connected to the network and use piped water as their main source of drinking water but report paying zero for their water. Anecdotal evidence supports the fact that many households are connected illegally to the network or that utilities do not invoice and collect revenues as they should.

For those households that report being connected to the piped water network but are located in areas for which we cannot map them to a specific provider, we use the median water quantity consumed for those we could impute the quantity (disaggregated by rural/urban and income deciles if enough observations are available).

Supplementary Appendix Table A1 shows the breakdown of the sample according to whether connected households are matched to corresponding volumetric tariffs and have positive water expenditures, allowing for the estimation of quantity consumed, or whether water quantities and the unit price paid have to be assumed using the information from other households in the data. The proportion of observations relying upon requiring the latter approach varies across countries significantly, with Nigeria and Bangladesh exhibiting the highest proportions.

Average unit price of water paid among those connected (UP). We define the unit price for a household h (UP_h) as the ratio of self-reported water expenditure from household surveys and quantity consumed. For a household with a water expenditure of zero, the unit price would also be zero.

Subsidy beneficiaries among those connected (B). We define variable B_h as equal to 1 if a household is connected and pays a unit price (UP_h) for water that is lower than the cost-reflective tariff. That is, a household is a subsidy beneficiary.

Value of the subsidy (S). The monthly value of the subsidy (S_h) is constructed using the variables described above; for each household, this is equal to $S_h = Q_{h|B} \times (CRT{-}UP_h)$, conditional on $UP_h < CRT$ (i.e., $B_h = 1$).

Rate of subsidization (R). The rate of subsidization for a household h, conditional on receiving a subsidy, is $R_{h|B} = 1{-}E_{h/B}/(Q_{h/B} \times CRT) = 1{-}UP_h/CRT$.

Distributional incidence of subsidies

Having identified beneficiary households (B_h) and the size of subsidies (S_h) they receive, we estimate the share of total subsidies accrued to each expenditure (or income) decile and consider whether they are progressive or not. This methodology is arithmetic, nonbehavioral, and expresses a partial equilibrium, which is well suited to investigate the incidence of subsidies at a point in time or with small marginal changes to prices or subsidies. This approach has limitations related to bigger changes in prices or subsidies. See, for instance, Abramovsky & Phillips (2015) for a description of various microsimulation models for tax and benefits using household survey data.

(1) *Progressive subsidies*: Subsidies are considered progressive when the share of subsidies accrued to each decile tends to decrease over the expenditure (or income) per capita distribution. This means that poor households capture a higher portion of the subsidy pie.
(2) *Regressive subsidies*: Conversely, subsidies are considered regressive when the share of subsidies accrued to each decile tends to increase over the expenditure (or income) distribution.

This relationship can sometimes be non-monotonic. In those cases, it is useful to compare the share of subsidies that the top decile receives relative to, for example, the bottom 40% of the distribution, which is known as the Palma ratio and widely used in developing countries (Cobham *et al.* 2016). Or the 20/20 ratio, the share of subsidies the top two deciles receive relative to the bottom 20% of the distribution. The higher the ratio, the more regressive the subsidies are.

We also estimate the share of overall countrywide household-level expenditure on goods and service captured by each decile to contrast with the share of piped water subsidy captured by each decile. This helps understand whether the subsidy is inequality reducing – that is, whether a greater percentage of subsidies is allocated to households in poorer deciles than is the case for overall countrywide expenditure. When the subsidy distribution is more skewed toward richer deciles than the distribution of expenditure or income, the subsidy is regressive and increases inequality. However, if subsidies are less regressive than the distribution of expenditure or income, they still can reduce inequality. These measures are simple statistics that can be easily displayed in bar charts and are conceptually similar to the quasi-Gini coefficients, described, for example, in Komives *et al.* (2005:140–41).

Targeting performance of subsidies and underlying mechanisms

Following Angel-Urdinola & Wodon (2007a), we define the targeting performance indicator (Ω), which relates to a subsidy's distributional incidence, as the ratio between the amount of subsidies the poor receive (S_P) and the amount of subsidies accrued to all households (S_H), divided by the proportion of poor households in the total population (P/H).

The targeting performance indicator (Ω) can be split into factors related to water network access and factors related to subsidy design, using the variables defined in the previous section and listed in Table 1:

$$\Omega = \frac{(S_P/P)}{(S_H/H)} = \frac{SA_P}{SA_H}\frac{C_{P|SA}}{C_{H|SA}}\frac{B_{P|C}}{B_{H|C}}\frac{R_{P|B}}{R_{H|B}}\frac{Q_{P|B}}{Q_{H|B}} \quad (1)$$

Access factors (SA and C) are fixed in the short term since they are determined by network expansion and households' decisions, and will affect the distributional incidence of the subsidy, regardless of the consumption subsidy's type or structure.

Two additional measures to assess the targeting performance of subsidies are the errors of inclusion and exclusion.

The error of exclusion (EE) is the share of households in poverty that is not benefiting from the subsidy: $EE = 1 - (B_P/P) = (P - B_P)/P$.

The error of inclusion (EI) is the share of beneficiary households that are not in poverty: $EI = B_{NP}/B_H = B_{NP}/(B_P + B_{NP})$. Figure 2 shows graphically how these errors are defined.

Descriptive statistics

Table 2 presents some descriptive statistics regarding household connection to piped water for each country analyzed.

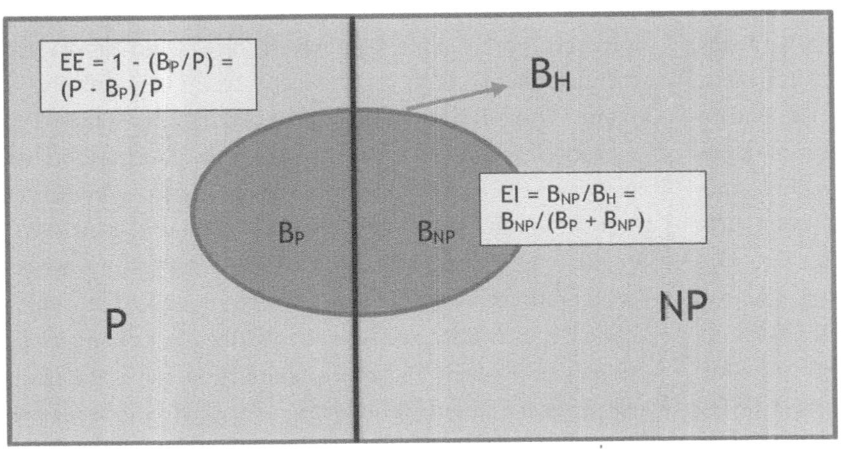

Figure 2 | Illustration of the errors of inclusion (EI) and exclusion (EE). *Note*: B_P, the number of targeted (beneficiary) households in poverty; B_{NP}, the number of targeted (beneficiary) households not in poverty; B_H, the number of total beneficiary or targeted households; P, the number of poor households; NP, the number of households not in poverty. The error of exclusion is $EE = 1 - (B_P/P) = (P - B_P)/P$. The error of inclusion is $EI = B_{NP}/B_H = B_{NP}/(B_P + B_{NP})$.

Table 2 | Households connected to piped water

Country (Year of household survey)	GDP per capita in USD PPP 2016 (1)	Households living in rural areas (%) (2)	Households connected to network water (piped water on plot) (%)			
			(3) Total	(4) Rural	(5) Urban	(6) Poor
Panama (2015)	29,446	30	91.9	79.3	97.3	82.6
Jamaica (2012)	9,551	47	69.7	49.0	88.9	57.4
El Salvador (2016)	8,288	36	79.3	64.1	88.0	67.9
Vietnam (2015)[a]	6,768	41	53.0	28.8	70.2	50.4
Nigeria (2016)	5,285	60	13.6	9.9	19.0	10.5
Bangladesh (2016)[b]	3,920	72	13.6	2.8	41.3	4.7
Mali (2014)	2,198	67	11.1	2.2	29.6	0.9
Ethiopia (2016)	1,896	73	19.8	3.5	63.9	2.4
Uganda (2014)	1,753	75	8.7	1.8	29.1	0.9
Niger (2014)	838	83	9.1	0.8	49.7	0.1

Source: Authors' own elaboration using several socio-economic household surveys, detailed in Supplementary Appendix Table A1, and World Bank Data series.
Note: These are countrywide figures. Poor households are defined as belonging to the first four deciles of the expenditure (or income) distribution in each country. All figures are calculated using sample weights. Water connection and public tap variables are derived from questions about the main source of drinking water – but some countries have information about the source of water for other uses (like Bangladesh), which are then also used to construct water connection. More information can be found in the Supplementary Appendix. Column 1 refers to GDP per capita, PPP (constant 2017 international U.S. dollars), GDP, gross domestic product; PPP, purchasing power parity (World Bank Data series, accessible at: https://data.worldbank.org/indicator/NY.GDP.PCAP.PP.KD).
[a]The figures for rural Vietnam using the 2015 Vietnam Household Registration Survey (VHRS) are much higher than the figures presented in JMP, using the Household Living Standards Survey (HLSS) 2012 that shows a rural rate of 13%. This is likely due to differences in the regional coverage of each survey. The VHRS 2015 has a sample that is representative of the population in five provinces – Ho Chi Minh City, Ha Noi, Da Nang, Binh Duong and Dak Nong – while the HLSS 2012 is supposed to be nationally representative.
[b]The figures for urban and total are higher than the figures published in JMP for the latest year. The data for Bangladesh used in this analysis differ from the data used in JMP. We use the Household Income and Expenditure Survey 2016, and the figures presented here are consistent with the official figures published in the Preliminary Report on Household Income and Expenditure Survey 2016, p. xiii (BBS 2016).

Countries are ordered by descending levels of GDP per capita (column 1). The figures reported in columns 2–6 are from household surveys and are, therefore, nationally representative, with the exception of Vietnam. (In Vietnam, the survey includes only five regions: Hanoi City, Da Nang, Dak Nong, Thu Dau Mot City and Bình Dương Province, and Ho Chi Min City.) Overall, richer countries tend to have a lower proportion of households living in rural areas and a higher proportion of households that report using piped water as their main source of drinking water. It is worth noting that there may be slight differences from the Joint Monitoring Program (JMP) data for each country, due to some minor differences in variable definitions in most cases and differences in sources in others. One common difference between our computation and JMP's computation is that we define a household as connected to piped water if it uses piped water as its main source of drinking water either in the dry or wet season. JMP also includes public standpipes as part of its piped water definition in some countries such as Ethiopia and Mali, whereas we include only house connections. Thus, our figures may be easily compared against JMP figures showing house connections only. For example, figures for Ethiopia and Nigeria are slightly higher in Table 2 (19.8% and 13.6%, respectively, in column 3) than those presented by JMP (15.2% and 12.2%, respectively). Despite wide variation in coverage across countries, all countries demonstrate higher connection levels in urban households than both rural and poor ones. As one might expect, there are also wide variations across the administrative geographic areas covered in the household surveys within each country, which are not shown in this paper. Notably, Nigeria's connection rates are quite low given the country's income per capita, but disparities among different population segments are smaller. Latin American countries have substantially higher connection rates, between 70 and 95% for the general population, and around 50% or over for rural and poor households. Vietnam falls in the middle, with a connection rate of 53% for the whole population and 70% for urban households.

Table 3 summarizes the number of providers from the IBNET dataset whose service areas are also covered in the household surveys, the average cost-reflective tariffs and imputed unit prices (conditional on the reporting of positive expenditure on water). More information about IBNET data for each country can be found in the Supplementary Appendix. Most countries have IBT structures, though two African countries – Nigeria and Uganda – have fixed rates. In Bangladesh, some service providers charge a flat water consumption rate, while others use IBT. El Salvador is the only country that has a volume-differentiated tariff (VDT) structure. Some countries or providers also add a value-added tax (VAT) on tariffs. For the analysis carried out in the following section, we use all the households included in every country's household survey, coupled with information from the corresponding utilities. As explained earlier in this section, when a household is located in an area or region with no available tariff structure, we use information from other households in that country to impute quantities and unit prices paid.

Remarkably, all countries show higher cost-reflective tariffs than the average unit price paid by all households that report paying a positive amount for piped water. This already shows that all countries' providers operate in a way that, on average, subsidizes residential consumers. However, 6 out of the 10 countries show a unit price higher than the estimated CRT_{opex}, suggesting that they can at least cover operating expenditures.

Limitations and strengths

The study presents limitations that must be taken into account when interpreting the findings, many of which

Table 3 | Tariff data and estimated unit costs and prices

Country	Number of providers covered in the analysis (1)	Type of tariff structure (2)	Estimated cost-reflective tariff (unit cost/m^3) CRT ($ 2017) (3)	Estimated OPEX cost-reflective tariff (unit cost/m^3) CRT_{opex} ($ 2017) (4)	Estimated effective average price/m^3 ($ 2017) (5)
Ethiopia	9	IBT	3.68	0.67	0.23
Mali	1	IBT with fixed charge + VAT	2.29	0.37	1.04
Niger	1	IBT	3.97	0.48	0.37
Nigeria	10	Fixed rate/m^3, some with a fixed rate	2.05	0.40	0.80
Uganda	1	Fixed rate/m^3	1.89	0.37	0.86
El Salvador	1	VDT, with fixed charge for 10 m^3 or less	0.85	0.19	0.32
Jamaica	1	IBT	2.51	0.55	1.69
Panama	1	IBT	1.18	0.26	0.20
Bangladesh	23	Depends on utility (IBT, 16 with flat rate)	0.49	0.12	0.26
Vietnam	3	IBT (+VAT for some)	2.70	0.62	0.38

Source: IBNET database, World Bank and other resources.
Note: Column 1: These are providers that covered areas included in the household surveys and for which there is available data online. Most of the data come from IBNET; however, for Nigeria, we use Abubakar (2016). *Ethiopia:* Data are from 2007 for Dire Dawa, and 2014 for Addis Ababa. The estimated cost-reflective tariff (column 3) is an average for the country cost-reflective tariff including both OPEX and CAPEX and is estimated in Andres *et al.* (2019), except for (a) Ethiopia (we use averages across the 24 Sub-Saharan African countries included in the estimation of Andres *et al.* (2019) – but excluding Ethiopia since the values for the cost-reflective tariffs for this particular country seem unrealistically high due to some problems with the raw data from Ethiopia used to estimate cost-reflective tariffs); (b) El Salvador (data from ANDA are used); (c) Panama (data from the Autoridad Nacional de Servicios Públicos are used for OPEX cost-reflective tariffs for the year 2015 in U.S. dollars and divided by 0.22, which is the relationship between the average of the OPEX cost-reflective tariffs across countries from Andres *et al.* (2019), and the average of total cost-reflective tariffs across countries from the same paper); and (d) Jamaica (National Water Commission). There are no available data on OPEX cost-reflective tariffs for El Salvador and Jamaica (column 4); thus, we use the value in column 3 for the total cost-reflective tariff, multiplying it by 0.22, which is the ratio between the average OPEX cost-reflective tariff and average total cost-reflective tariff from Andres *et al.* (2019). Column 5 shows the unit price conditional on paying for water services (i.e., self-reported expenditure on water greater than zero). The figure is calculated using sampling weights. CRT, cost-reflective tariff; CRT_{opex}, OPEX cost-reflective tariff; IBT, increasing block tariff; m^3, cubic meter; CPEX, operating expenditure; VAT, value-added tax; VDT, volume-differentiated tariff.

have already been discussed. The study also has important strengths. We summarize them below.

Due to data constraints, the data used spans different periods within each country. For example, household data for Jamaica is from 2012, whereas the tariff data is from 2013. Or within Nigeria, we have data on tariff structures for nine regional providers, spanning from 2012 to 2018, while the household data is from 2015 to 2016.

The data on water expenditure are self-reported and can sometimes include expenditure not only on piped water but also other water sources. If the piped water supply is shared with other households, water expenditure figures may not accurately reflect the amount paid by the reporting household. Since this variable is used to impute the quantity of water consumed and the average unit price paid by households, these variables can suffer from inaccuracies. For example, if the expenditure is overestimated, then quantities will be overestimated. If the measurement issue varies with the level of income of households, this could be biasing our results for those households that are connected to the network. In addition, in many countries, a significant proportion of households report missing or zero expenditure on water even if they report using piped water on premises as their main source of drinking water. Quantities for these households must be imputed from the other households within the sample.

In some countries, we only have data on tariff structures for some service areas, which means that, for a significant proportion of connected households, we must use average prices and quantities consumed from those areas with data. We have explained how we impute quantities in the section 'Quantity of water consumed among those connected (Q)', and we have discussed that this limitation is more important in countries like Nigeria and Bangladesh. This could affect the distributional impact of subsidies across deciles, the calculation of omega and the calculation of inclusion and exclusion errors, but it is not clear in which direction. Having said this, the main driver of regressivity is the lack of access or connection. The unavailability of data on expenditure or tariffs would not affect these measures. However, they could affect the magnitude of the regressivity conditional on being connected.

The definition of the service area may be too large and inaccurate, and this may overestimate the actual number of households with potential access to the piped water network. This may affect the calculation of omega and underestimates the magnitude of the impact of access factors on targeting performance. It will not, however, affect the distributional impact analysis of subsidies nor the calculation of the errors of inclusion or exclusion. Additionally, we are comparing countries with varying proportions of rural populations and differing degrees to which piped water is a feasible solution to close the access gap.

One of the main strengths of this study is that it uses representative samples that cover the poorest households; these households are least likely to have access and be connected to piped water. Studies that restrict their sample to only those households for which water quantity data are available to fail to be representative at the country level. If the objective of policymakers is to use limited public funds to help the poorest access safely managed water, then the strengths of our approach outweigh its limitations.

FINDINGS

Distributional performance of existing subsidies

In this section, we show the estimated overall distributional incidence of water consumption subsidies across households by expenditure decile (or income decile for El Salvador and Panama).

As explained in Section 'Distributional incidence of subsidies', each figure shows the percentage of subsidy captured by each decile (i.e., percentage of money spent on subsidies accruing to all households in the given decile, classifying all households in the country along the countrywide expenditure distribution). In addition, we present the share of total household income or expenditure on goods and services in the economy accruing to each expenditure decile and compare it with the share of subsidies accruing to each decile. When the subsidy distribution is more skewed toward richer deciles than the distribution of expenditure or income, it is regressive and increases inequality. If the subsidy shares are less skewed toward richer deciles that total expenditure, but still regressive, they can be regressive but with inequality reducing.

Presented in Figure 3, the estimates for the five African countries demonstrate a common trend: richer households

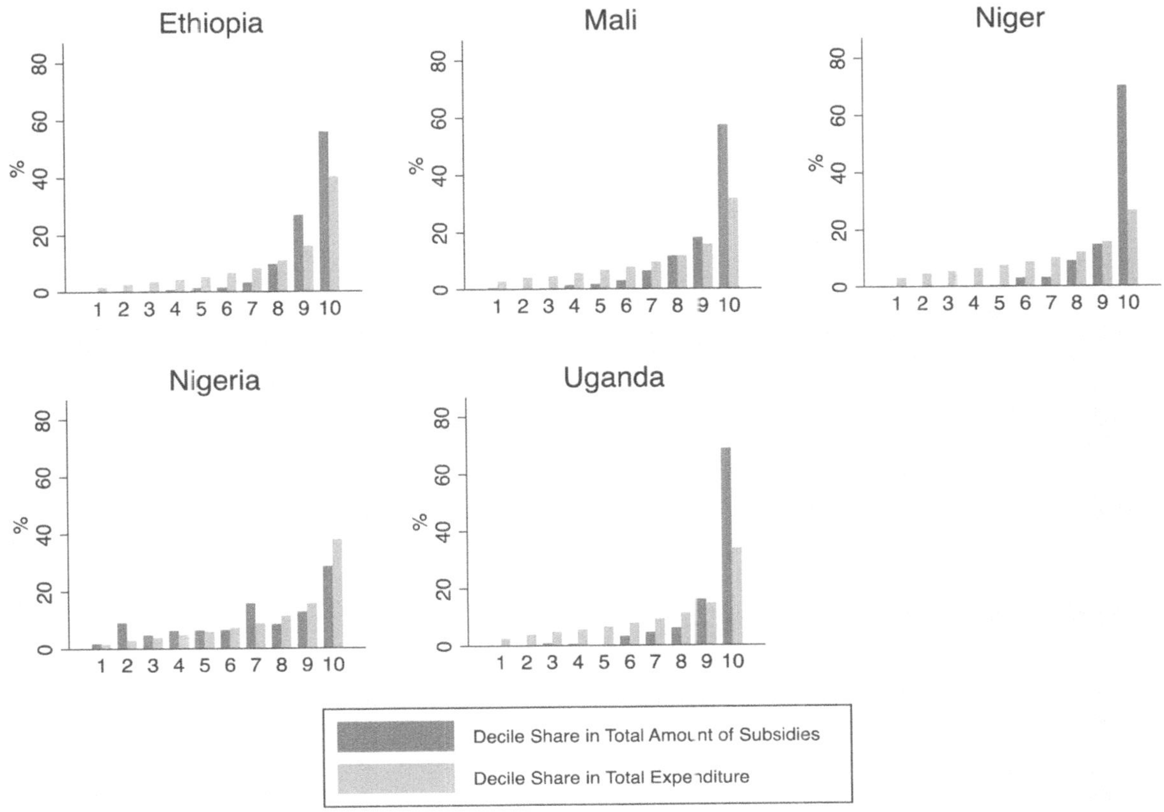

Figure 3 | Distributional incidence of subsidies (by decile) for five African countries. *Source:* Authors' own elaboration based on household surveys, IBNET and administrative data on tariff structure, and cost-reflective tariff data. See the section on Methodology and Data and the Supplementary Appendix for more detail. *Note:* All figures are calculated using sample weights. Total expenditure is the total household expenditure in all categories of goods and services. The distribution of expenditure refers to the countrywide distribution of expenditure per capita, i.e., households are ranked according to their expenditure per capita.

appear to enjoy a greater share of subsidies. In Niger and Uganda, households in the top decile are estimated to receive over 60% of all water consumption subsidies. In Mali, Niger and Uganda, the estimates show that the poorest hardly benefit from the subsidy at all. Furthermore, Nigeria seems to be the only country where subsidies reduce rather than increase inequality according to these estimates. Table 4 further shows the summary measures of distributional incidence. It shows that the subsidies are most regressive in Niger, with the top 20% of households receiving almost 85% of subsidies and the bottom poorest households receiving 0%.

Figure 4 shows the estimated distribution of beneficiaries (households that receive the subsidy) across expenditure deciles for the same five countries to complement the picture painted by distributional incidence. This figure helps to visualize the poor performance of these subsidies in targeting the poor by presenting the number of beneficiaries rather than the value of subsidies received across the deciles of total expenditure. Higher deciles appear to show the highest share of total subsidy beneficiaries (dark gray bars) and the highest share of all households within each decile that are beneficiaries (light gray bars). The differences between the top three deciles and the remaining ones are remarkable in all countries, although less pronounced in Nigeria. As noted above, in Niger, the poor (first four deciles) do not seem to benefit from the subsidy at all; Uganda shows a similar pattern, and in Mali, only the fourth decile shows a very small number of beneficiaries. We conclude that in the five African countries analyzed, and using the household data, tariff data and the methodology described, consumption subsidies are regressive (albeit relatively less so in Nigeria) and not well targeted to the poor. This seems to be due to

Table 4 | Measures of distributional incidence of subsidies

	Percentages of subsidies accruing to each group of deciles of the overall total expenditure distribution					
	Bottom 20%	**Bottom 40%**	**Top 20%**	**Top 10%**	**Palma ratio (Top 10%/Bottom 40%)**	**20/20 Ratio (Top 20%/Bottom 20%)**
Ethiopia	0.54	1.56	82.60	55.76	35.71	153.87
Mali	0.65	2.26	75.13	57.17	25.24	115.39
Niger	0.00	0.12	84.54	70.00	577.43	22,315.38
Nigeria	10.82	21.79	41.15	28.44	1.30	3.81
Uganda	1.27	1.27	84.76	68.69	54.16	65.83
Average in five African countries	*2.66*	*5.40*	*73.64*	*56.01*	*138.77*	4,531.06
El Salvador	6.03	17.81	37.88	21.85	1.23	6.28
Jamaica	7.46	19.70	41.78	24.65	1.25	5.60
Panama	12.44	27.29	32.66	17.67	0.65	2.62
Average in three Latin American countries	*8.64*	*21.60*	*37.44*	*21.39*	*1.04*	*4.84*
Bangladesh	4.57	12.36	52.70	32.28	2.61	11.54
Vietnam	12.62	30.30	26.19	14.20	0.47	2.07
Average in two Asian countries	8.60	21.33	39.45	23.24	1.54	6.81
Average across all countries	5.64	13.45	55.94	39.07	70.01	2,268.34

Source: Authors' own elaboration based on household surveys, IBNET and administrative data on tariff structure, and cost-reflective tariff data. See the section on Methodology and Data and the Supplementary Appendix for more detail.

Note: All figures are calculated using sample weights. Total expenditure is the total household expenditure in all categories of goods and services. The distribution of expenditure refers to the countrywide distribution of expenditure per capita, i.e., households are ranked according to their expenditure per capita.

the poor appearing to have less access to piped water and, even when connected, appearing to consume less water generally as shown in Table 5 and discussed in more detail in the section 'Targeting performance of existing subsidies: are subsidies pro-poor?'.

Presented in Figure 5, the results for Panama, Jamaica and El Salvador have a similar pattern, demonstrating subsidies' regressivity. However, the figures suggest that subsidy schemes in all three Latin American countries reduce inequality since subsidy incidence is less skewed toward richer deciles than total expenditure/income. The degree of subsidy regressivity in Latin American countries is lower than in African countries, as evidenced by the average Palma and 20:20 ratios shown in Table 4. As explored in Section 'Targeting performance of existing subsidies: are subsidies pro-poor?', this is partly because both a greater number of poor households reside in service areas, and that, conditional on this potential access, a greater percentage of poor households are connected to the network.

Figure 6 shows the distribution of beneficiaries (households that receive the subsidy) across expenditure or income deciles for the three countries. As with the African countries, it complements the picture painted by the distributional incidence of the amount of subsidies. Higher deciles show the largest share of total subsidy beneficiaries (dark gray bars) and the largest share of all households within each decile that are beneficiaries (light gray bars), although in Panama the pattern is less pronounced.

Figure 7 shows the results for two Asian countries, Bangladesh and Vietnam. In Bangladesh, water consumption subsidies are strongly regressive and increase inequality. While subsidies in Vietnam are still regressive, albeit less so, they are actually inequality reducing, suggesting slightly better targeting compared to Bangladesh. Figure 8, which depicts the distribution of beneficiary households across deciles, shows strong regressivity in Bangladesh but a more even distribution across expenditure deciles in Vietnam, once again suggesting better targeting.

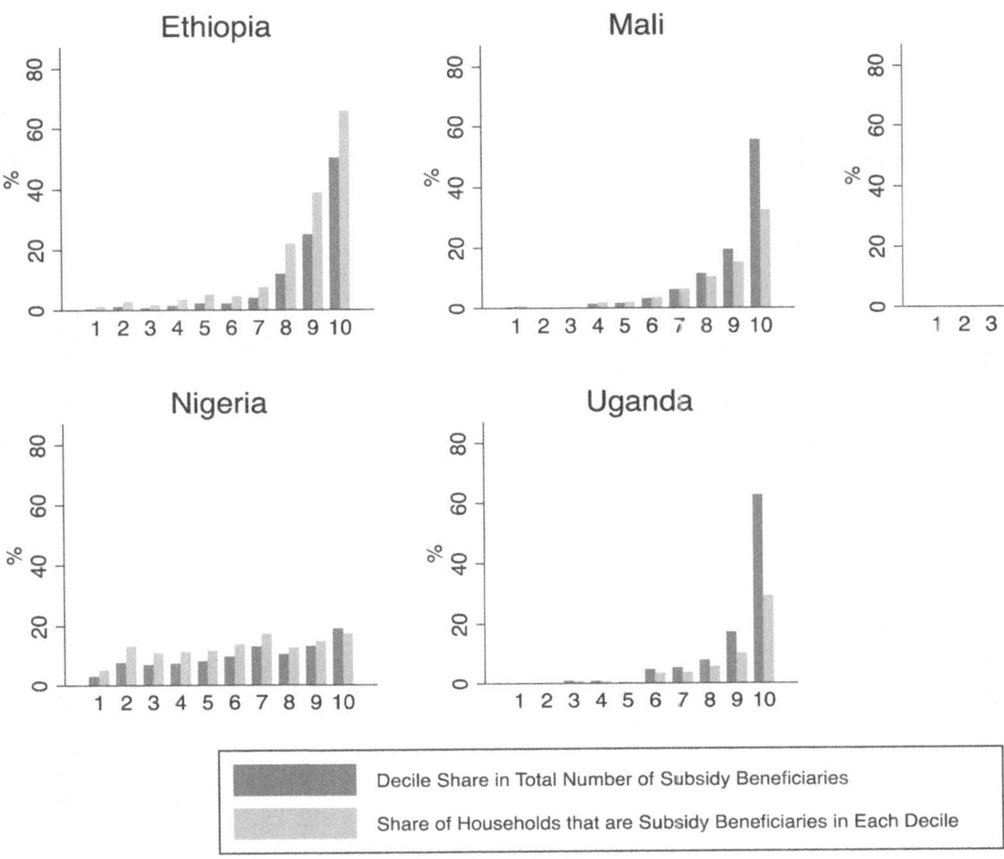

Figure 4 | Distributional incidence of subsidy beneficiaries (by decile) for five African countries. *Source:* Authors' own elaboration based on household surveys, IBNET and administrative data on tariff structure, and cost-reflective tariff data. See the section on Methodology and Data and the Supplementary Appendix for more detail. *Note:* Subsidy beneficiaries are households. All figures are calculated using sample weights. Total expenditure is the total household expenditure in all categories. The distribution of expenditure refers to the countrywide distribution of expenditure per capita, i.e., households are ranked according to their expenditure per capita.

In the next section, we look at a more synthetic measure of targeting performance that complements this analysis and consider both access and subsidy design factors driving this performance.

Targeting performance of existing subsidies: are subsidies pro-poor?

Figure 9 and Table 5 show the targeting performance of water consumption subsidies in the 10 countries under analysis. Figure 9 shows how access factors ($SA*C$) and subsidy design factors ($B*R$) drive omega (Ω) across countries. Table 5 presents the values of each factor for each country. Factor values below 1 indicate that the factor contributes to reduced targeting performance, whereas factor values above 1 indicate that the factor contributes to increased performance.

As shown in Figure 9, all African, Latin American and Asian countries present Ω values below 1, demonstrating that water consumption subsidies are poorly targeted to the most vulnerable populations. Access factors (dark gray dots) are the driving force explaining low Ω values in most countries. Subsidy design factors (light gray dots) are, in general, around or above 1. These indicators suggest that government improvement strategies should mainly focus on increasing access to services (for each country, we explore whether strategies should primarily focus on expanding service areas, increasing the proportion of

Table 5 | Decomposing the targeting performance of water consumption subsidies

	Ω (1) Ω > 1 → pro-poor	SA (2) Potential access to water connection (service area)	C (3) Connection rate (for those with access)	B (4) Receipt of subsidy (for those connected)	R (5) Rate of subsidization (for those with a subsidy)	Q (6) Quantities (for those connected)
Ethiopia	0.047					
Poor		0.255	0.095	1.000	0.957	3.275
All		0.446	0.444	0.996	0.964	8.567
Poor/All		0.571	0.214	1.004	0.993	0.382
Mali	0.085					
Poor		0.036	0.263	0.907	0.858	14.491
All		0.239	0.467	0.958	0.793	14.751
Poor/All		0.150	0.563	0.948	1.081	0.982
Niger	0.006					
Poor		0.030	0.040	1.000	0.921	10.453
All		0.209	0.433	1.000	0.909	23.651
Poor/All		0.141	0.093	1.000	1.014	0.442
Nigeria	0.661					
Poor		0.507	0.208	0.997	0.929	3.270
All		0.585	0.233	0.993	0.900	3.972
Poor/All		0.867	0.894	1.004	1.032	0.823
Uganda	0.079					
Poor		0.204	0.045	1.000	0.544	5.173
All		0.332	0.261	1.000	0.549	6.965
Poor/All		0.610	0.17	1.00	0.99	0.740
El Salvador	0.839					
Poor		0.945	0.719	0.996	0.679	19.230
All		0.966	0.821	0.983	0.647	20.863
Poor/All		0.977	0.876	1.013	1.050	0.922
Jamaica	0.718					
Poor		0.998	0.575	0.967	0.764	10.643
All		0.998	0.698	0.955	0.653	14.456
Poor/All		1.000	0.823	1.013	1.170	0.736
Panama	0.859					
Poor		1.000	0.826	1.000	0.872	40.497
All		1.000	0.919	1.000	0.829	44.518
Poor/All		1.000	0.898	1.000	1.052	0.910
Bangladesh	0.345					
Poor		0.286	0.164	0.998	0.999	36.555
All		0.384	0.355	0.998	0.995	36.579
Poor/All		0.745	0.462	1.000	1.003	0.999

(continued)

Table 5 | continued

	Ω (1) Ω > 1 → pro-poor	**SA (2) Potential access to water connection (service area)**	***C* (3) Connection rate (for those with access)**	***B* (4) Receipt of subsidy (for those connected)**	***R* (5) Rate of subsidization (for those with a subsidy)**	***Q* (6) Quantities (for those connected)**
Vietnam	0.858					
Poor		0.804	0.627	1.000	0.861	13.863
All		0.840	0.631	1.000	0.858	15.430
Poor/All		0.958	0.993	1.000	1.004	0.898

Source: Authors' own calculations using country-specific household surveys, administrative data and estimated cost-reflective tariffs.
Note: Access factors: $SA = SA_P/SA_H$ is the share of households located in a service area for the poor relative to the population as a whole; $C = C_{P|SA}/C_{H|SA}$ is the relative share of households that are connected to the service conditional on potential access (being located in a service area). Subsidy design factors: $B = B_{P|C}/B_{H|C}$ is the relative proportion of households with a subsidy conditional on usage; $R = R_{P|B}/R_{H|B}$ is the relative average rate of subsidization; $Q = Q_{P|B}/Q_{H|B}$ are relative average quantities consumed. A factor value below 1 indicates that the factor contributes to reduced targeting performance, whereas a value above 1 indicates that the factor contributes to increased targeting performance. All figures are calculated using sample weights. Poor households are defined as belonging to the first four deciles of the expenditure (or income) distribution in each country.

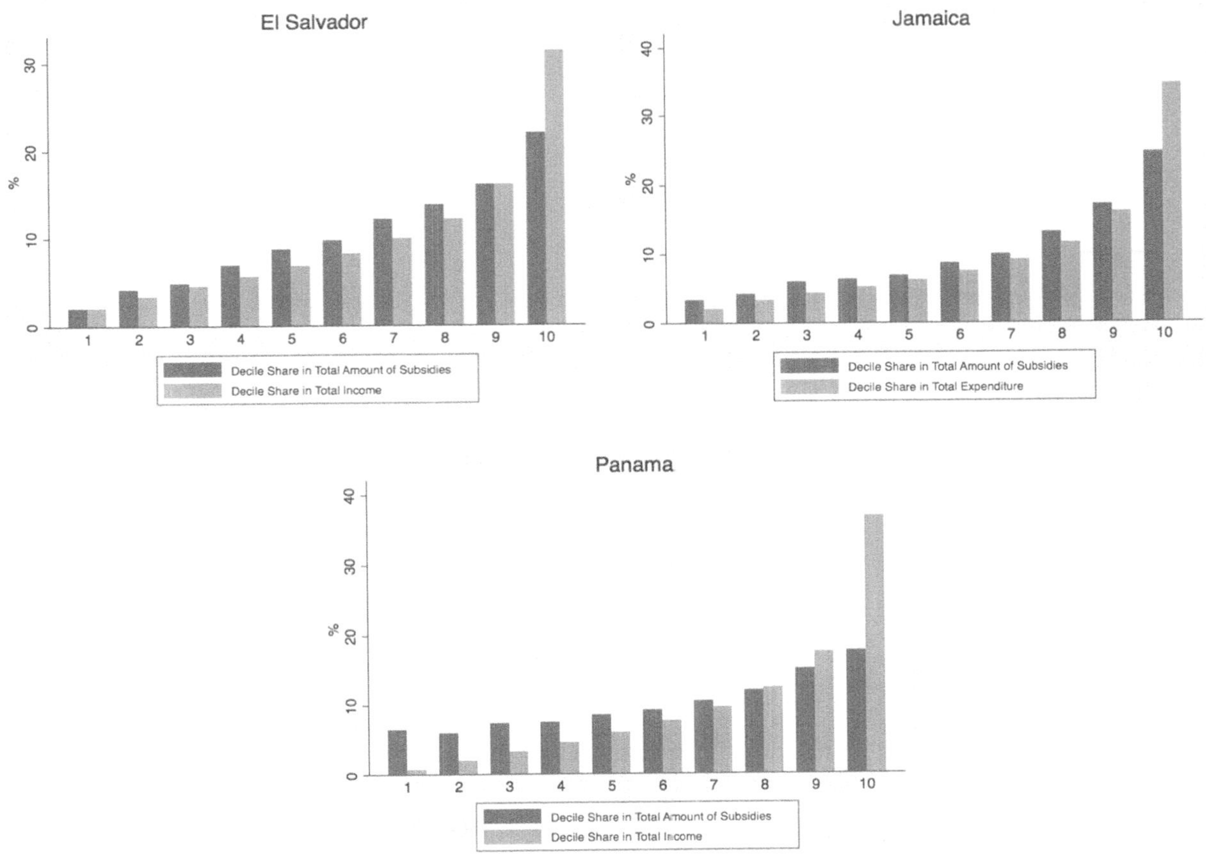

Figure 5 | Distributional incidence of subsidies (by decile) for three Latin American countries. *Source:* Authors' own elaboration based on household surveys, administrative data and cost-reflective tariff data. See the section on Methodology and Data and the Supplementary Appendix for more detail. *Note:* All figures are calculated using sample weights. Total expenditure is the total household expenditure in all categories. The distribution of expenditure (Jamaica) or income (El Salvador and Panama) refers to the countrywide distribution of expenditure/income per capita, i.e., households are ranked according to their expenditure/income per capita.

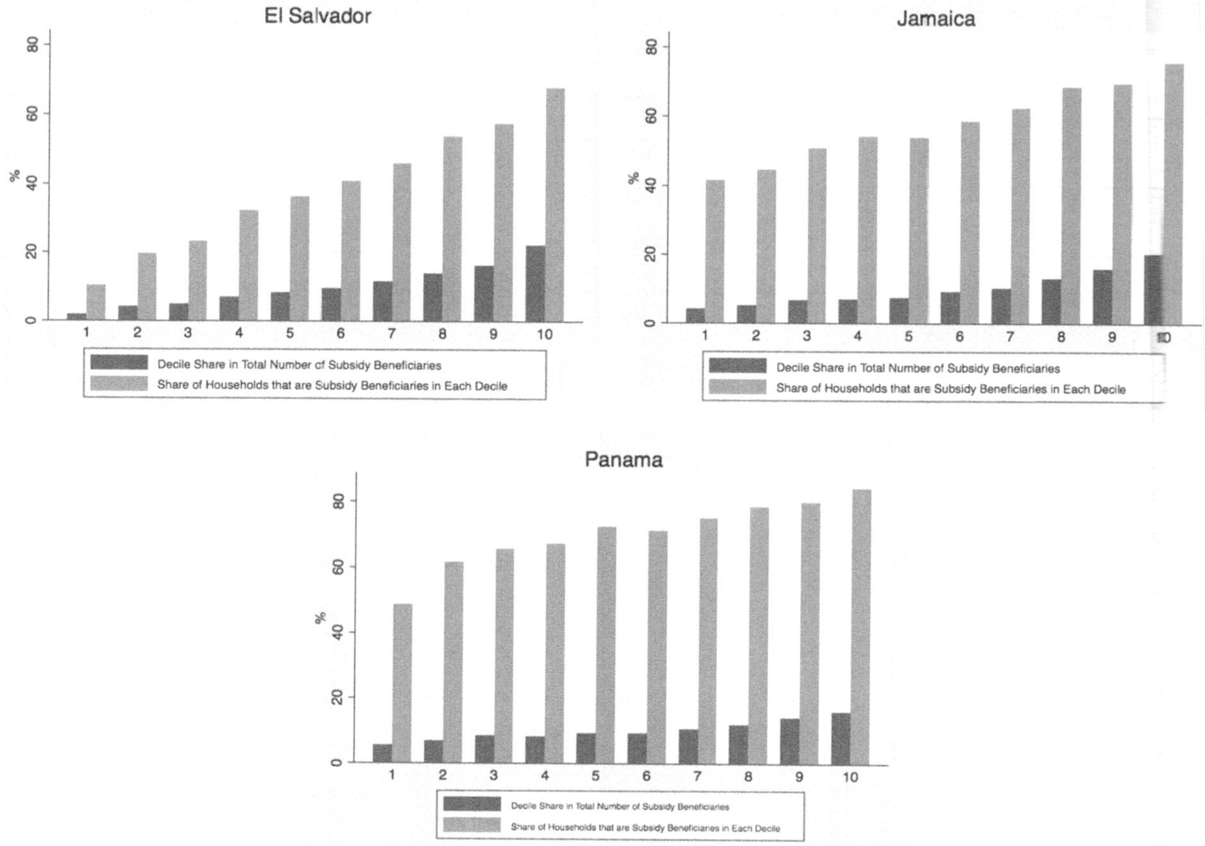

Figure 6 | Distributional incidence of subsidy beneficiaries (by decile) for three Latin American countries. *Source:* Authors' own elaboration based on household surveys, administrative data and cost-reflective tariff data. See the section on Methodology and Data and the Supplementary Appendix for more detail. *Note:* All figures are calculated using sample weights. Total expenditure is the total household expenditure in all categories. The distribution of expenditure (Jamaica) or income (El Salvador and Panama) refers to the countrywide distribution of expenditure/income per capita, i.e., households are ranked according to their expenditure/income per capita.

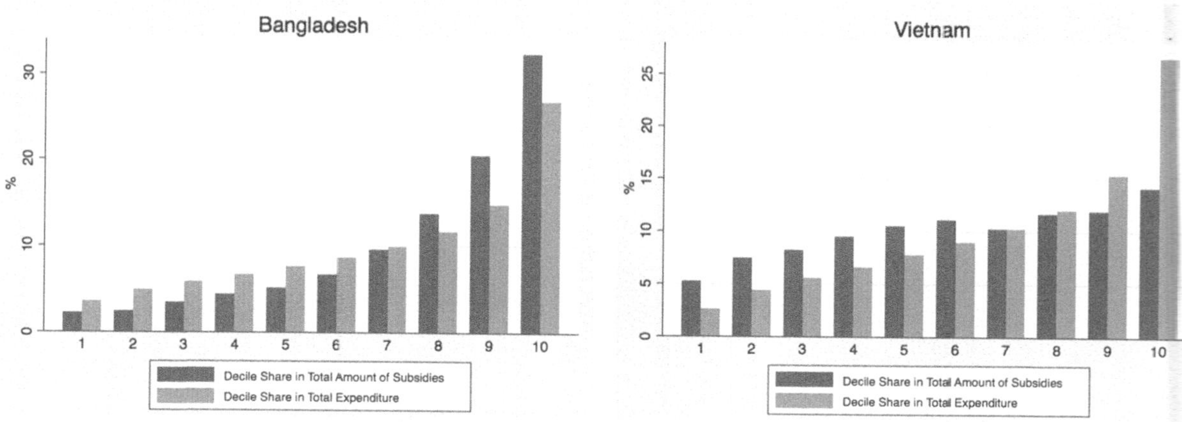

Figure 7 | Distributional incidence of subsidies (incidence by deciles) for two Asian countries. *Source:* Authors' own elaboration based on household surveys, administrative data and cost-reflective tariff data. See the section on Methodology and Data and the Supplementary Appendix for more detail. *Note:* All figures are calculated using sample weights. Total expenditure is the total household expenditure in all categories. The distribution of expenditure refers to the countrywide distribution of expenditure per capita, i.e., households are ranked according to their expenditure per capita.

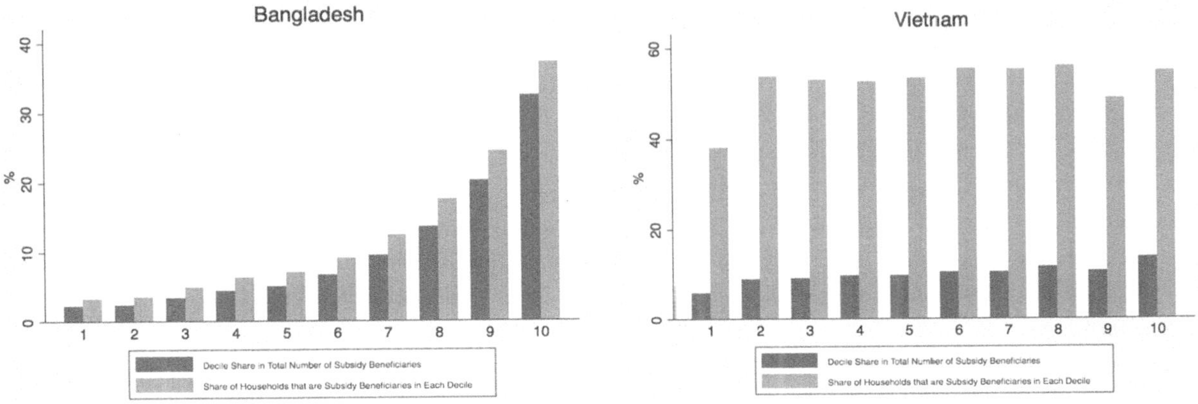

Figure 8 | Distributional incidence of subsidy beneficiaries (incidence by deciles) for two Asian countries. *Source:* Authors' own elaboration based on household surveys, administrative data and cost-reflective tariff data. See the section on Methodology and Data and the Supplementary Appendix for more detail. *Note:* Subsidy beneficiaries are households. All figures are calculated using sample weights. Total expenditure is the total household expenditure in all categories. The distribution of expenditure (Jamaica) or income (El Salvador and Panama) refers to the countrywide distribution of expenditure/income per capita, i.e., households are ranked according to their expenditure/income per capita.

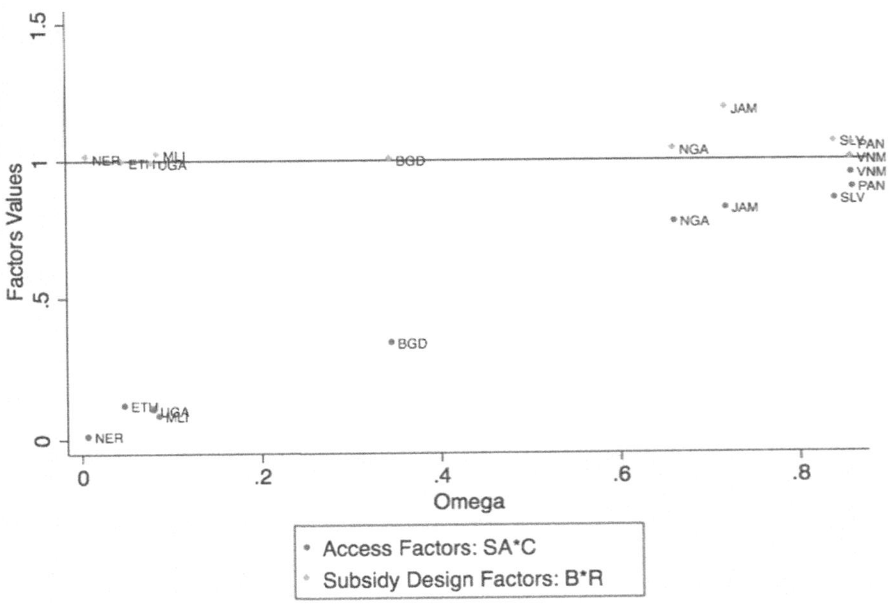

Figure 9 | Factors driving the targeting performance of subsidies (Ω) across countries. *Source:* Authors' own elaboration based on household surveys, administrative data and cost-reflective tariff data. See the section on Methodology and Data and the Supplementary Appendix for more detail. *Notes:* Country labels: BGD: Bangladesh, ETH: Ethiopia, JAM: Jamaica, MLI: Mali, NER: Niger, NGA: Nigeria, UGA: Uganda, PAN: Panama, SLV: El Salvador, VNM: Vietnam. Access factors: $SA = SA_P/SA_H$ is the share of households located in a service area for the poor relative to the population as a whole; $C = C_{P|SA}/C_{H|SA}$ is the relative share of households that are connected to the service conditional on potential access (being located in a service area). Subsidy design factors: $B = B_{P|C}/B_{H|C}$ is the relative proportion of households with a subsidy conditional on usage; $R = R_{P|B}/R_{H|B}$ is the relative average rate of subsidization. Factor values below 1 indicate that the factor contributes to reduced targeting performance, whereas factor values above 1 indicate that the factor contributes to increased targeting performance.

connected households within existing service areas, or both in detail for each country), although enhancing subsidy design could also be beneficial.

Different factors influence the severity of poor subsidy targeting performance across countries. In Table 5, we explore the different access and subsidy design factors in more detail.

Four African countries – Ethiopia, Mali, Niger and Uganda – show an estimated value of Ω lower than 0.1. In all four, the main factors that reduce subsidy efficiency in targeting poor households relative to all households include (i) poor households' lower probability of being located in a service area and (ii) their lower probability of being connected to a water network conditional on being located in a service area. Additionally, in Ethiopia and Niger, the relatively low water quantities consumed by the poor significantly decrease the targeting performance.

Nigeria shows an Ω above 0.5, revealing a significant problem, although less acute than in Ethiopia, Mali, Niger and Uganda. SA, C and Q are closer to 1, and $B = 1$ and $R = 1.03$. Consistent with the evidence in the previous section, Nigeria's distributional incidence of subsidies by decile is regressive, although less pronounced than in the other four African countries.

In order to improve the targeting performance of water consumption subsidies, these governments should expand the service area (i.e., SA) of their utilities and connect more households within this service area (i.e., C) to the water network. One way to do this is by implementing connection subsidies in areas where there is already network infrastructure present (Andres *et al.* 2019 discuss different types of connection subsidies and beneficiaries and the relevant considerations when thinking of how to design them depending on the context). Supplementary Appendix Table A2 shows the number of households that are not connected to the network that uses a public tap, which could be connected to the existing infrastructure. This varies significantly across countries and across households classified as rural, urban and poor. Such considerations are important when planning connection subsidies for a specific context. Additionally, these countries should improve the design of their water consumption subsidies to better target poor households (increasing B to values over 1).

In El Salvador, Jamaica and Panama, the problem is less acute, with an Ω of 0.84, 0.69 and 0.86, respectively. In these countries, the factors lowering Ω to below 1 are different. In all three, the probability of living in a service area is high for both types of households, poor and non-poor (SA is close to 1). However, connection rates conditional on location in a service area and water quantities consumed are both significantly lower for poor households, driving Ω to below 1. These countries would benefit from increasing service connection rates, as well as improving subsidy design to better target the poor (resulting in an increase of variables B and R further above 1).

Bangladesh falls in between African and Latin American countries, with an Ω of 0.35. Both the variables SA and C are low in general, and lower for poor households. Water quantities consumed by the poor are similar to those of other households, but this is potentially driven by the lack of information on households' expenditure and the flat rates being charged, which complicates imputing Bangladesh's household-specific water consumption (therefore, a median consumption quantity is assumed for most households). A parameter Q close to 1 is potentially reasonable given that any connected household regardless of consumption pays a flat rate. Nonetheless, potentially all factors (SA, C, B and R), in addition to the tariff structure, could be improved. In Vietnam, the value for Ω is much higher, though still below 1, and a focus on subsidy design (B and R) could increase targeting performance.

Table 6 shows the errors of inclusion (the proportion of subsidy beneficiary households that are in the top six deciles, or rich) and the error of exclusion (the proportion of

Table 6 | Water consumption subsidies, errors of inclusion and exclusion

	Error of inclusion (%) (1)	Error of exclusion (%) (2)
Ethiopia	96.01	97.57
Mali	97.82	99.15
Niger	99.73	99.93
Nigeria	74.35	89.59
Uganda	98.28	99.60
El Salvador	82.01	78.27
Jamaica	51.78	51.78
Panama	71.20	38.91
Bangladesh	87.67	95.45
Vietnam	66.51	50.28

Source: Authors' own calculations using country-specific household surveys, administrative data and estimated cost-reflective tariffs.
Note: Poor households are defined as belonging to the first four deciles of the expenditure (or income) distribution in each country. Error of inclusion is measured by the percentage of all beneficiary households that are rich; the error of exclusion is measured by the percentage of poor households that do not get a subsidy. All figures are calculated using sample weights.

poor households that do not receive a water consumption subsidy). These errors are not conditional on location in a service area or connection, so they differ from those in Table 5. The estimates show that for most countries covered here the errors of exclusion range from 78% to nearly 100%, except for Jamaica, Panama and Vietnam (column 2). This is consistent with the very low levels of Ω presented in Table 5. The countries on the lower end of this spectrum exhibit errors of exclusion around 50%, suggesting that their level of outreach to the poor could still be improved.

Inclusion errors (column 1 of Table 6) are extremely high in all countries, suggesting a wide margin for the improvement of targeting, consistent with the analysis presented so far. The inclusion errors are particularly high in African countries, with figures close to 100% in all countries besides Nigeria. Bangladesh also stands out with figures close to 90%.

SUMMARY AND DISCUSSION

This paper provides new evidence on the recent performance of piped water consumption subsidies in terms of pro-poor targeting for 10 low- and middle-income countries around the world.

We use detailed household survey data with information on whether the households' main source of drinking water is from piped water to define access, and the households' monthly expenditure on water services combined with available information on tariff structures in each country to impute water quantities and unit prices paid by each household. We supplement this information with new estimates of the cost incurred by the supplier of providing piped water to households to calculate the average subsidy per unit of water consumed at the household level.

Our accounting methodology presents some limitations, most importantly due to the lack of information on water quantity consumed at the household level and the tariff structure faced by each household in each country. This means that water quantities have to be imputed. Furthermore, the tariff structures for some households have to be assumed based upon suppliers operating in other regions within the same country. Finally, we are comparing countries with varying levels of urbanization and rural contexts. As a result, piped water is more feasible to close the access gap in some countries than in others.

With these caveats in mind, our analysis of 10 low- and middle-income countries suggests that piped water is substantially subsidized, as existing tariffs do not seem to fully recover the costs of service provision. Moreover, these consumption subsidies do not appear to efficiently address service gaps among the poor due to ineffective targeting.

Overall, we find that water consumption subsidies in all 10 countries are regressive and, therefore, do not adequately target the poor (the estimated Ω has a value under 1 in all cases, with an average value of 0.45 across all 10 countries). This is despite their high diversity of economic development, national income levels, proportion of rural populations, piped water coverage and the costs of producing and distributing piped water. We also determine that the severity of regressivity varies significantly (from 0.006 in Niger to 0.87 in Panama). On average, across the 10 low and middle-income countries examined, 56% of subsidies end up in the pockets of the richest 20%, but only 6% of subsidies find their way to the poorest 20%.

Only a very small number of poor households benefit from subsidies, particularly within the African countries in our sample. This is because poor households are less likely to be located in areas serviced by utility providers (the variable SA in the analysis), and even where they are, they are less likely to be connected and consuming piped water (variable C) than the general population.

This suggests that access factors primarily drive this poor targeting performance (SA takes a value of 0.7 and C takes a value of 0.6, on average, across all 10 countries). This problem is particularly pressing in the African countries (with an average SA of 0.47 and average C of 0.39 across the sample used) and Bangladesh (which has an SA of 0.75 but a C of 0.46). This is expected since these countries have large rural populations, aggravating the feasibility of extending the network to new service areas. However, connection rates are far from 100% even in urban areas in these countries (Table 2), implying that there is also scope to improve connection rates. Furthermore, in Ethiopia, Mali and Niger, available data show that there is a significant proportion of households that are not connected to the network and that use a public tap as the main source of drinking water, both in urban and rural areas (Supplementary Appendix Table A2). This could indicate the potential feasibility of increasing connection rates for these households.

Aside from improving access for the poor, subsidy design could also be improved to better target the poor. The share of poor households that receive a subsidy conditional on being connected to the network relative to the equivalent share for the whole population (B) averages 0.99 and the rate of subsidization conditional on receiving a subsidy for the poor relative to the equivalent rate for the whole population (R) averages at 1.02, across all 10 countries. This signals a need for improvement in subsidy design even given subpar network coverage and connection rates. This trend is observed in all countries.

It is worth noting that there is not an optimal value of omega; it is a policy choice that will vary across countries according to their particular fairness preferences and political equilibriums. If the objective is to use available and limited public funds to ensure poor households can access and consume piped water in an affordable and sustainable way, then an omega below 1 demonstrates that this objective is not being achieved.

To summarize, our results show that in most developing countries, improving the targeting of current water consumption subsidy schemes to the poorest households will primarily require improving their access to the service. This is because, by definition, unconnected households are excluded from the pool of subsidy beneficiaries. Therefore, network expansion into poorer neighborhoods, if technologically feasible and policies that facilitate household connections should be pursued. Angel-Urdinola & Wodon (2012) provide evidence from Nicaragua that increased access rates among the poor over time improves the targeting performance of water consumption subsidies.

It is also important to note that the targeting performance of connection subsidies would also depend on the context, design of the subsidy and the behavior of both utilities and households. For example, universal connection subsidies were simulated by Komives *et al.* (2005), assuming that all unconnected households were offered and accepted a subsidized connection. In countries where the proportion of unconnected households is higher among the poor than among the overall population, subsidies are likely to be pro-poor. Yet, in practice, poor households may be located in areas more difficult to reach or face additional financial or technical barriers to connection, such as the inability to afford the fixtures required to connect to the network, preventing them from benefiting from connection subsidies (Komives *et al.* 2006).

In countries where water access is no longer a significant issue and connection rates are relatively high, such as El Salvador, Jamaica and Panama, simply modifying a subsidy's design could greatly improve its targeting of the poor. In some countries, quantity-based subsidies are used in combination with subsidies targeted using administrative selection. But in other countries, only administrative selection, like geographic targeting or mean testing, is used. In general, administrative selection performs better than quantity-based subsidies in targeting the poor (Komives *et al.* 2006). Having said this, the error of exclusion can be quite high since in an effort to target the poor accurately, a significant proportion of poor households are excluded.

Further detailed analysis of each context would be needed to design the most cost-effective set of policy instruments for improving the pro-poor performance of consumption subsidies, including those that improve access to safely managed water among the poor. The most desirable approach will vary across, and even within, countries. It will depend on state capacity to implement more refined and explicit mechanisms to target the poor, such as administrative selection, as well as whether the poor live in areas with an existing water network, whether they are able to undertake the necessary upgrades within the home for connection, and whether network expansion, if required, is feasible.

SUPPLEMENTARY MATERIAL

The Supplementary Material for this paper is available online at https://dx.doi.org/10.2166/washdev.2020.073.

REFERENCES

Abramovsky, L. & Phillips, D. 2015 *A Tax Micro-Simulator for Mexico (MEXTAX) and Its Application to the 2010 Tax Reforms*. IFS Working Paper W15/23. Institute for Fiscal Studies, London.

Abubakar, I. R. 2016 Quality dimensions of public water services in Abuja, Nigeria. *Utilities Policy* **38**, 43–51.

Andres, L. A., Thibert, M., Lombana Cordoba, C., Danilenko, A. V., Joseph, G. & Borja-Vega, C. 2019 *Doing More with Less: Smarter Subsidies for Water Supply and Sanitation*. World Bank, Washington, DC. © World Bank: https://openknowledge.worldbank.org/handle/10986/32277. License: CC BY 3.0 IGO.

Angel-Urdinola, D. & Wodon, Q. 2007a Do utility subsidies reach the poor? Framework and evidence for Cape Verde, Sao Tome, and Rwanda. *Economics Bulletin* **9** (4), 1–7.

Angel-Urdinola, D. & Wodon, Q. 2007b Does Increasing Access to Infrastructure Services Improve the Targeting Performance of Water Subsidies? Available from: https://ssrn.com/abstract=1133125 or http://dx.doi.org/10.2139/ssrn.1133125.

Angel-Urdinola, D. & Wodon, Q. 2012 Does increasing access to infrastructure services improve the targeting performance of water subsidies? *Journal of International Development* **24** (1), 88–101.

Baietti, A. & Curiel, P. 2005 *Financing Water Supply and Sanitation Investments. Estimating Revenue Requirements and Financial Sustainability*. Water Supply and Sanitation Working Notes. Note No.7, October 2005. World Bank Group.

BBS (Bangladesh Bureau of Statistics) 2016 *Preliminary Report on Household Income and Expenditure Survey*. BBS, Dhaka. Available from: http://bbs.portal.gov.bd/sites/default/files/files/bbs.portal.gov.bd/page/b343a8b4_956b_45ca_872f_4cf9b2f1a6e0/HIES%20Preliminary%20Report%202016.pdf.

Bitrán, G. & Arellano, P. 2005 Regulating water services: sending the right signals to utilities in Chile (English). Public Policy for the Private Sector; Note no. 286. World Bank Group, Washington DC. Available from: http://documents.worldbank.org/curated/en/619981468237324751/Regulating-water-services-sending-the-right-signals-to-utilities-in-Chile.

Cardenas, H. & Whittington, D. 2019 Magnitude and Distribution of Electricity and Water Subsidies for Households in Addis Ababa, Ethiopia. Policy Research Working Paper; No. 9025. World Bank, Washington, DC.

Chen, M., Guixia, F., Lidan, W., Zhonghua, W., Yuxin, Z. Y. & Lei, S. 2015 Who benefits from government healthcare subsidies? An assessment of the equity of healthcare benefits distribution in China. *PLoS ONE* **10** (3), e0119840.

Cobham, A., Schlögl, L. & Sumner, A. 2016 Inequality and the tails: the Palma proposition and ratio. *Global Policy* **7** (1), 25–36.

Deaton, A. 2019 *The Analysis of Household Surveys (Reissue Edition with a New Preface): A Microeconometric Approach to Development Policy*. World Bank, Washington DC.

Ebeke, C. H. & Ngouana, C. A. L. 2015 *Energy Subsidies and Public Social Spending: Theory and Evidence*. International Monetary Fund. Working Paper WP/15/101.

Ebinger, J. O. 2004 *Measuring Financial Performance in Infrastructure: An Application to Europe and Central Asia*. World Bank. Working Paper Series WPS3992.

Goksu, A., Trémolet, S., Kolker, J. & Kingdom, B. 2017 *Easing the Transition to Commercial Finance for Sustainable Water and Sanitation*. World Bank, Washington, DC. Available from: http://hdl.handle.net/10986/27948.

GWI (Global Water Intelligence) 2004 *Tariffs: Half Way There*. GWI, Oxford, UK

Hounsa, M. T., Coulibaly, M. & Sanoh, A. 2019 *The Redistributive Effects of Fiscal Policy in Mali and Niger*. Poverty and Equity Global Practice, Working Paper Series No. 208. World Bank Group, Washington, DC. Available from: http://documents.worldbank.org/curated/en/966741560877543325/The-Redistributive-Effects-of-Fiscal-Policy-in-Mali-and-Niger.

Hutton, G. & Chase, C. 2017 Water supply, sanitation, and hygiene. In: *Disease Control Priorities (3rd edn): Vol. 7, Injury Prevention and Environmental Health* (C. N. Mock, R. Nugent, O. Kobusingye & K. Smith, eds). World Bank, Washington, DC.

Hutton, G. & Varughese, M. C. 2016 *The Costs of Meeting the 2030 Sustainable Development Goal Targets on Drinking Water Sanitation, and Hygiene* (English). Water and Sanitation Program Technical Paper. World Bank Group, Washington, DC. Available from http://documents.worldbank.org/curated/en/415441467988938343/The-costs-of-meeting-the-2030-sustainable-development-goal-targets-on-drinking-water-sanitation-and-hygiene (accessed 2 February 2018).

Inchauste, G. & Lustig, N. 2017 *The Distributional Impact of Taxes and Transfers: Evidence From Eight Developing Countries. Directions in Development – Poverty*. World Bank, Washington, DC.

Kochhar, K., Pattillo, C. A., Sun, Y. M., Suphaphiphat, N., Swiston, A. J., Tchaidze, R., Clements, B. J., Fabrizio, S., Flamini, V., Redifer, L. & Finger, H. 2015 *Is the Glass Half Empty Or Half Full?: Issues in Managing Water Challenges and Policy Instruments*. International Monetary Fund. Staff Discussion Notes No. 15/11.

Komives, K., Foster, V., Halpern, J., Wodon, Q. & Abdullah, R. 2005 *Water, Electricity, and the Poor: Who Benefits From Utility Subsidies?* (English). Directions in Development/World Bank, Washington, DC. Available from: http://documents.worldbank.org/curated/en/606521468136796984/Water-electricity-and-the-poor-who-benefits-from-utility-subsidies.

Komives, K., Halpern, J., Foster, V., Wodon, Q. & Abdullah, R. 2006 *The Distributional Incidence of Residential Water and Electricity Subsidies*. Policy Research Working Paper No. WPS 3878. World Bank, Washington, DC.

O'Donnell, O., van Doorslaer, E., Wagstaff, A. & Lindelow, M. 2008 *Analyzing Health Equity Using Household Survey Data: A Guide to Techniques and Their Implementation*. World Bank, Washington, DC.

Whittington, D., Nauges, C., Fuente, D. & Du, X. 2015 A diagnostic tool for estimating the incidence of subsidies delivered by water utilities in low- and medium-income countries, with illustrative simulations. *Utilities Policy* **34**, 70–81.

WHO/UNICEF 2017 *Progress on Drinking Water, Sanitation and Hygiene: 2017 Update and SDG Baselines*. Licence: CC BY-NC-SA 3.0 IGO. World Health Organization (WHO) and the United Nations Children's Fund (UNICEF), Geneva.

World Bank Group 2017 *Reducing Inequalities in Water Supply, Sanitation, and Hygiene in the Era of the Sustainable Development Goals: Synthesis Report of the WASH Poverty Diagnostic Initiative*. World Bank, Washington, DC. Available from: https://openknowledge.worldbank.org/handle/10986/27831.

WWAP (UNESCO World Water Assessment Programme) 2019 *The United Nations World Water Development Report 2019: Leaving No One Behind*. UNESCO, Paris.

WWAP (United Nations World Water Assessment Programme) 2016 *The United Nations World Water Development Report 2016: Water and Jobs*. UNESCO, Paris.

First received 25 March 2020; accepted in revised form 23 June 2020. Available online 20 July 2020

Research Paper

Does payment by results work? Lessons from a multi-country WASH programme

Guy Howard and Zach White

ABSTRACT

Payment by results (PbR) for financing public services has attracted increasing interest over recent years in the water, sanitation, and hygiene (WASH) sector. PbR is attractive to funders as a mechanism because it focuses attention on results rather than inputs, and because it transfers a proportion of risk to suppliers. This paper reviews the experience of the UK Department for International Development (DFID) funded WASH Results Programme (WRP), which used PbR, drawing on a process evaluation and the experience of the first author in commissioning the programme, and the second author in evaluating it. The WRP met its targets for people reached with first-time access to water and sanitation and generated high-quality programme data. The PbR mechanism provided strong incentives to the suppliers to improve their monitoring systems. However, the suppliers tended to use tried and tested approaches, with limited innovation. It is critical to consider certain key elements in the design of PbR programmes, including the proportion of funding that uses PbR and the proportions of PbR that focuses on outputs and outcomes.

Key words | payment-by-results, finance, results-based finance, WASH results, WASH targets

Guy Howard (corresponding author)
Department of Civil Engineering,
University of Bristol, University Walk,
Bristol BS8 1TR,
UK
E-mail: *guy.howard@bristol.ac.uk*

Zach White
Water, Sanitation, and Hygiene Team,
Oxford Policy Management,
Clarendon House, Level 3, 52 Cornmarket Street,
Oxford, OX1 3HJ,
UK

INTRODUCTION

Payment by results (PbR) is an approach to financing public services that has gained increasing attention over recent years (Cabinet Office 2011). PbR is a form of outcome-based performance management that provides financial rewards for the achievement of specific outcomes (Albertson *et al.* 2018a). PbR and other approaches to outcome-based performance management are derived from the sustained efforts of governments, particularly in high-income countries, to improve efficiency and effectiveness in service provision as part of reforming public services, which has been referred to as the 'New Public Management' model (Hood 1991).

The use of PbR is attractive to funders of public services because it places a focus on achieving a desired set of results, rather than managing a range of inputs. It transfers a portion of the delivery risk to suppliers, which is accompanied by increased responsibility of suppliers for delivery (National Audit Office 2015). In a review of the drivers for developing PbR, Albertson *et al.* (2018b) note that there may be several objectives for PbR, including expansion of the supplier market and diversifying the supplier base; improving efficiency in public service delivery; and managing complex issues with multiple factors that influence outcomes. Clist (2017) notes that the key aspect for PbR is the quality of the performance measure and that in order to create the right incentives such measures must effectively capture a result that all those involved in a project really care about. In the context of water, sanitation, and hygiene (WASH), this includes expanded access to safe drinking water and/or sustained use of services.

From the beginning of 2010, the UK Government placed an increasing focus on expanding the use of PbR in ODA-spending

doi: 10.2166/washdev.2020.039

through the Department for International Development (DFID). By the end of 2013, 71% of centrally issued DFID contracts had a performance-based element. In 2014, DFID published its strategy on PbR, which included commitments to make PbR its 'business as usual' approach in contracts with its suppliers, rather than the exception (DFID 2014b).

The evidence base from evaluations of PbR programmes in the context of official development assistance (ODA) remains limited. Clist (2017) and Fox & Morris (2019) note that there remain few evaluations of PbR programmes and that the quality of evaluations is generally poor. Albertson *et al.* (2018b) summarised that the findings of the evaluations that were available, primarily based on examples from the UK. They found that PbR has often associated with more complex commissioning and increased commissioning costs; mixed results in expanding supplier markets; an increased focus on performance, often through strengthened monitoring, which had both positive and negative effects; increases in delivery costs in some cases; no evidence of extensive innovation and that suppliers were risk-averse; and limited evidence of 'gaming' the PbR system.

PbR programmes can deploy downside and upside incentives to the suppliers contracted. Downside incentives mean that failure to meet expected targets results in a loss of payment. Upside incentives reward suppliers with additional payments should the desired results be exceeded. Clist (2017) concluded that a key element for the success of PbR is the ability and willingness of the funder to withhold payment, especially in relation to non-governmental organisations (NGOs). This focus on downside incentives, however, tends to make suppliers risk-averse and to stifle innovation.

This paper reviews the experience of one WASH programme that used PbR – the WASH Results Programme (WRP) – to examine how PbR has worked in practice, and the lessons learnt. The paper examines the key question of whether the programme delivered the expected results for the funder of the programme. The paper does not assess whether PbR worked *better* than other approaches to funding WASH. There are insufficient data to make a fair comparison as data were not collected from comparable programmes that had the same set of target results, similar geographies and similar timeframes, which would be required to make a comparative analysis.

THE WASH RESULTS PROGRAMME

The WRP is a £111 million programme running from 2013 to 2022. It was originally designed to meet a UK Government target for numbers of people gaining access to WASH between 2010 and 2015, but further extended in 2016 to support the achievement of greater numbers of people gaining first-time access to WASH between 2015 and 2020.

The WRP is delivered through three supplier contracts: the South Asia WASH results programme (SAWRP) consortium, led by Plan International; the Sustainable WASH in Fragile Contexts (SWIFT) consortium, led by Oxfam; and the SNV Sustainable Sanitation and Hygiene for All (SSH4A) programme. The contracts include two distinct sets of results associated with two phases of activity: an output phase, focused on ensuring first-time access to sanitation and/or water supply supported with hygiene education and a subsequent outcome phase to maintain use of sanitation and/or water supplies constructed and to maintain good hygiene practice.

Table 1 summarises the suppliers' programmes and the number of people expected to be reached. The WRP achieved, and in most cases exceeded, the results targets established. At the end of the output phase under the original contracts, the WRP projects had provided over 1 million people with first-time access to water supply, over 4 million people with first-time access to sanitation, and over 10 million people reached with hygiene messages (DFID, 2016). The outcome phase focused on continued use of services as shown in Table 1.

At the start of the programme, an independent team of sector experts were contracted to provide verification that the results claimed by suppliers had been achieved, and therefore, payments could be made. Verification was primarily based on systems appraisals, assessing whether supplier monitoring and data systems were robust and reliable and, therefore, producing credible data, with limited spot-checks in the field. An autonomous team undertook both process and impact evaluations.

The WRP only used downside incentives, operating on a sliding scale down to 70% of the target, after which full payment would be lost. The WRP was designed as

Table 1 | Details of supplier targets and countries included in programmes and outcome results from the original contracts (DFID 2018)

Supplier	Total target people to be reached	Countries	Outcome results
SAWRP	2,279,761	Bangladesh, Pakistan[a]	Bangladesh: 96.1% of people continued to use latrines and 95% continued to use water supplies Pakistan: 99% of people continued to use latrines and 100% continued to use water supplies
SSH4A	3,471,797	Bangladesh, Nepal, Ethiopia, Ghana[a], Kenya, Mozambique, South Sudan[b], Uganda, Zambia	2.85 million people continued to use latrines (exceeded target by 30%)
SWIFT	1,684,402	Democratic Republic of Congo (DRC), Kenya[a], Liberia[b]	70% of people continued to use latrines and 75% of people continued to use water supplies

[a]Country not included in contract extension.
[b]Country where activities were stopped early.

'100% PbR' – that is, all payments were contingent on the achievement of a pre-defined set of results. However, payments were not solely linked to output- and outcome-level 'results': interim results were set within each phase to allow payments to be made throughout the programme. Many early payment triggers were essential programme activities and inputs (e.g. training workshops).

METHODS

This paper draws on data collected through a process evaluation of the WRP (ePact 2018) and complements this with an analysis of the experience of the first author in the commissioning and management of the programme. We draw these two forms of evidence together through a critical analysis with greater weight given to empirical data over expert experience. The analysis of the findings of the evaluation and expert opinion are structured to primarily address the needs of the funder, as the data collection was designed to meet this purpose.

The data collected through the evaluation included: key informant interviews with DFID management, the global management of the three supplier consortia; interviews with programme management in the 11 WRP countries; and case studies in 4 of the 11 countries, which included interviews with field-level staff, government counterparts, and beneficiaries. The primary data were collected over the course of the evaluation, supplemented by a literature review, a review of programme documentation, and reviews of the programme annual reviews and business cases.

This evaluation drew on elements of contribution analysis and realist evaluation to assess the degree to which the PbR modality influenced implementation. This paper focuses on one of the core hypotheses explored in the evaluation. Namely, that the introduction of a PbR modality helped to achieve intended outputs and outcomes. This assertion is based on three related propositions:

1. the programme and its PbR modality allowed the flexibility of implementation approach within the sub-programmes, which helped to achieve output and outcome objectives;
2. stronger monitoring systems as a result of the PbR modality increased the likelihood of achieving intended outputs and outcomes; and
3. the results-oriented problem-solving promoted under the PbR modality increased the likelihood of achieving intended outputs and outcomes.

The analysis presented here does not cover the full breadth of the evaluation but provides a more detailed assessment on these three critical propositions and explores key issues of interest to funders seeking to use PbR. This includes the degree to which the PbR modality helped to achieve DFID's stated market-shaping objectives.

There are limitations to the study presented here, most importantly that the data collected and analysed does not permit a comparative assessment of PbR performance for the reasons noted above, and that detailed value for money analysis was not possible. In addition, the focus on the needs of the funder, means that the experience and impact on suppliers and communities, is not fully captured.

RESULTS

Did the PbR modality allow the flexibility of implementation approach within the sub-programmes, which in turn helped to achieve output and outcome objectives?

While there was flexibility in how suppliers met their targets, there is little evidence of innovation. There were several important reasons for this, including that many of the design decisions were made by partners before the PbR modality was fully understood. Furthermore, in the face of high downside delivery risks, the partners generally adopted tried and tested approaches in contexts familiar to them, as these could predictably deliver results. The main flexibility that supported the achievement of the targets was the ability of the suppliers to use multiple projects to deliver results, meaning that shortfalls in one country could be offset by achievements in other countries.

The evaluation found that the removal of financial and activity reporting requirements did result in programme managers being able to more flexibly manage programmes, but that this flexibility tended to stay at the higher levels of programme management. There were two principal reasons for this. First, many consortium leads were unwilling to transfer that level of risk to partners as they did not believe they had the ability to pre-finance activities. Secondly, there was a perceived need to tightly manage partners and field teams to ensure results were delivered, and complete financial autonomy at that lower level was seen as too risky.

Did stronger monitoring systems as a result of the PbR modality increase the likelihood of achieving intended outputs and outcomes?

The systems-based approach to verification under the WRP provided significant incentives to the suppliers to improve their monitoring systems and the suppliers made substantial investments to improve the reliability and robustness of their monitoring systems, which led to more robust data. These improvements were sustained in the 'outcome phase' following the delivery of outputs. The evaluation found that the suppliers underestimated the amount of time, effort, and resources required to upgrade their monitoring. Given that PbR was unfamiliar to suppliers and the DFID team, the evaluation concluded that an inception phase would have been beneficial to allow for the design of verification before implementation. The WRP suppliers and verifiers iterated successfully around verification processes, but this was inefficient and could have been avoided by having greater clarity earlier in the programme.

Verification of the more complex aspects of programming (e.g. sustainability prerequisites or learning) lacked agreed sector-wide standards for measurement proved difficult to measure. Furthermore, establishing appropriate outcome-level targets was challenging because attribution was more difficult, and because of the limited evidence of suitable benchmarks for the conversion of WASH outputs into outcomes.

Did the results-oriented problem-solving promoted under the PbR modality increase the likelihood of achieving intended outputs and outcomes?

The achievement of results targets within the expected timeframe, which was challenging both in terms of numbers of people and limited implementation time, demonstrated that the PbR modality using strong performance measures, created positive incentives for a focus on results delivery. However, the evaluation of the WRP found that the approach of linking all payments to results was 'neither necessary nor optimal'. The application of PbR in the early stages of the programme meant the results being verified were process-related indicators primarily used to facilitate cash flow, rather than to incentivise supplier attention to specific aspects of programming. This was inefficient as these indicators were costly and time-consuming to document and verify.

Over the course of the programme, three external shocks severely impacted on programme results. In each case, these events led to negotiations surrounding the appropriate level of supplier–DFID risk-sharing as the results could not be fully achieved as originally intended. These events were the Ebola outbreak in West Africa in 2014, a resurgence of violence in South Sudan in 2016, and a prolonged drought in East Africa in 2017–2018. In each of these cases, DFID *de facto* assumed some of the delivery risk *post hoc*. In the case of Ebola, the contract was renegotiated to exclude affected countries, with adjustments upwards in price per beneficiary made to compensate for the sunk costs of establishing the programme in the cancelled countries. In the case of South Sudan, the results from that country were permitted to be made up through progress in other countries. In the case of the drought in East Africa, an allowance was made in the payment decision after the verification team determined that the severity of the drought was unusual and 'could not have been expected during the period of the WASH Results Programme', and recommended the full payment be made despite the full outcome-level results not being achieved. Negotiations around these events were time-consuming and required the suppliers, verifiers, and DFID to carry out additional assessments.

Did the use of the PbR modality contribute to DFID market-shaping objectives?

One of DFID's key aims in using PbR for the WRP was to build the supplier base and to attract new market entrants. The WRP did identify NGOs capable of managing WASH programmes at a scale comparable to the United Nations Children's Fund, which provides one of the best benchmarks for larger-scale donor-funded WASH programmes. However, given the limited number of NGOs able to demonstrate the ability to manage such large projects, market-shaping was limited. All the successful lead suppliers were organisations that had either significant WASH programmes in the proposed countries, allowing them to better manage financial risks, or were established consortia. No private sector organisation led a successful bid under the WRP, and, overall, the private sector response to this opportunity was more limited than had been anticipated by DFID.

One of the challenges in commissioning PbR projects is the availability of reliable price benchmarks for tender evaluation, which was a problem at the onset of the WRP. As the suppliers worked across a wide range of contexts, it was challenging to assess what were reasonable variations in price caused by the delivery location, the particular groups targeted, and other qualitative factors.

DISCUSSION

The lessons from the WRP raise some interesting questions about the application of PbR in WASH programmes, and the use of PbR more broadly. The WRP experience suggests that the greatest value of PbR was in focusing supplier attention on specific aspects of programme implementation.

Flexibility in implementation

A key theorised benefit of PbR is that it allows for greater innovation (DFID 2014a, 2014b). This assertion is rooted in the idea that because suppliers are only accountable for results and are not tied to reporting against activity-based workplans, they are able to propose novel approaches and have greater flexibility in implementation. Similar to the findings of Albertson *et al.* (2018b) and the broader emerging evidence base, the WRP experience indicates that this does not occur in practice. PbR tends to promote risk-averse behaviour by suppliers, who stick to tried and tested approaches and locations in which they have greater confidence in their ability to deliver expected results, and so secure payments. It is possible that the strict and pressured time limits for achieving results, the limited experience in the use of PbR and the structure of the contracts reinforced this behaviour. Funders seeking to support the development of genuinely novel and innovative approaches must acknowledge the need for a 'space to fail'. In such cases, where 'failure' is a very real possibility of the financial risks associated with PbR appear likely to become unacceptable for both funder and supplier.

Stronger monitoring and verification

System-based verification reinforced incentives for suppliers to improve their monitoring but potentially came at the cost of securing 'hard' independent data. This was a trade-off that was supported by the spot-checking of results by the verifiers to provide confidence that monitoring systems were reliable and robust. The use of spot-checks allowed issues of concern to be raised and discussed at payment decisions meetings, and in some cases resulted in deferred payments until such time as suppliers could provide reliable data. The alternative of verification through primary independent data collection would have significantly raised costs to the funder and would have reduced the incentives for suppliers to make their own reporting more robust.

The limited inception phase created issues for the efficiency of the monitoring. A longer inception phase would have allowed the more complex elements of verification of systems and in particular outcomes to be addressed. Although the suppliers underestimated the additional costs of improving monitoring systems, for the WRP, in the view of the authors, the benefits from improved quality of data from better monitoring systems and strong technical verification justified the additional costs. However, it is noted that these benefits were largely confined to improvements within the NGO partners, and there was limited impact on government monitoring practices and systems in the countries in which the suppliers were working.

Impact on outputs and outcomes

The WRP experience confirms the findings of Clist (2017) that PbR works best when it is focused on issues of shared concern between supplier and funder and that selecting the right performance measures is critical to the success of a PbR programme. However, linking payment to the early 'results' packages comprised of activities was inefficient to verify and of little benefit to improving programming. This suggests that future programmes using PbR should consider using a mixed model of grant and PbR funding, applying the former to essential inputs and the latter to the delivery of target output results. Identifying strong performance measures at the outcome level proved more difficult than identifying those at the output level, which was partly because outcome measures were not agreed at the outset of the contracts. In future PbR programmes, greater attention should be given to resolving these questions early in design, and before programming starts.

In designing effective PbR WASH programmes, careful consideration is needed regarding the proportion of total funding that uses PbR, and what proportion is linked to output and outcomes. Higher-risk PbR contracts (those with a large proportion of payment specified on outcomes) may raise the risk profile to unacceptably high levels for suppliers, who may, therefore, choose not to bid on contracts. Conversely, if the PbR element is too small, it may fail to offer real incentives. In terms of longer-term sustainability, PbR may be less suited for programmes focused on system-building, as the indicators that currently exist for measuring systems development are related to processes and activities rather than outcomes. They rely on qualitative evidence and subjective judgement, making them challenging and expensive to verify. Furthermore, unless the PbR contract is with a government, the supplier will have very little direct control over the results.

This suggests that PbR is not suited for use in all WASH programmes; its use should be governed by the specific rationale for using PbR, and in instances where reliable and verifiable outcome indicators are available. Where PbR is applied at both the output and outcome level, as in the WRP, it is important to consider how these are linked. For SAWRP and SWIFT, the outcome targets were defined as proportions of the contracted output-level beneficiaries rather than as a proportion of the actual output-level beneficiaries reached. As such, where output targets were substantially exceeded, as they were in many countries, the outcome targets became substantially less demanding. In future PbR programmes, it would be preferable to base outcome targets on the proportion of the actual output population reached. The experience of the WRP in the face of unexpected events – Ebola in West Africa, Violence in South Sudan, and drought in East Africa – highlighted that PbR afforded no greater benefit in being able to operate in fragile contexts and in each of the cases mentioned that the programme was either suspended or allowances were made in payment decisions. The WRP findings suggest that, overall, PbR is likely to be better suited to non-fragile contexts and if the application of PbR in fragile states is pursued, it is likely that more extensive risk assessment would be required at the design stage.

The WRP only used downside incentives, though there are arguments for the use of upside incentives in PbR programmes from a supplier perspective (it lowers the risk profile while maintaining a focus on results), to motivate suppliers to achieve more results than planned, or if the purpose of the PbR modality is to incentivise improved quality in implementation. However, upside incentives can pose challenges for funders, who need predictability in multi-year budgets, especially when these funders are government agencies. Donor financial systems are often better at accommodating uncertainty associated with downside incentives. For instance, previous behaviour can be modelled to estimate whether suppliers have a record of under-spend, and this can be used as the basis for budget projections. This permits the reallocation of available funds on a month-by-month basis. Upside incentives may create greater uncertainties for the funder, which are propagated into the future and have an impact on medium-term budget profiles. This tends to have a more significant impact on budget accuracy and planning, and as such while upside incentives may be preferred by suppliers, and offer different incentives to purely downside incentives, they can be challenging to execute for some donors.

Market shaping

Similar to the findings of Albertson *et al.* (2018b), market shaping under the WRP gave somewhat mixed results. New large contracts were established, but the overall number of NGOs realistically able to manage such large, time-sensitive contracts was small. It is unclear whether NGOs willing to bid for such large contracts would do so in a context in which they would be moving into a country without pre-existing programmes. This suggests that non-geographically specific programmes may benefit from a wider market, but programmes focused on individual countries may not. It is noteworthy that no private sector organisation led a successful bid under the WRP, and, overall, the private sector response to this opportunity was more limited than had been anticipated by DFID.

CONCLUSION

The WRP has shown that PbR can be an effective financing instrument for rural basic WASH projects. It has delivered significant results at the output level and shown that these can be sustained for at least 2 years post-implementation at high levels. PbR was effective in promoting investment in monitoring systems by suppliers and independent verification increased the quality and reliability of the results. As with the previous use of PbR, it is essential that strong performance measures are agreed that address issues that both suppliers and funders care about. The experience under the WRP highlights that the PbR modality did not stimulate innovation and, in fact, promoted risk-averse behaviour by suppliers. The WRP experience also indicates that future WASH programmes using PbR will need to incorporate an inception phase to agree performance measures and processes of verification, and to obtain clarity on outcomes. It is also recommended that future programmes only apply PbR to a portion of total funding reserved for incentivising specific aspects of programming and that essential project inputs are funded through grant mechanisms. There is a need for further evaluations of PbR in practice to establish where, how, and in what ways this modality can be most effectively used, and in the coming years to synthesise the emerging evidence from the PbR programmes that have finished and been evaluated. The question of whether PbR should be used in preference to other modalities remains unanswered and will depend on the context of the programme design. Where there are clear targets and no expectation of innovation, then PbR is well suited. Where objectives are more complex or where innovation is desired, other financing mechanisms may be more appropriate.

ACKNOWLEDGEMENTS

This work draws on the evaluation funded by DFID, primarily under the contract Monitoring, Verification and Evaluation Service Provider for the WASH Results Programme. The views expressed in this paper are solely those of the authors, they do not represent the position of DFID or OPM. The authors would like to acknowledge the wider group of authors responsible for the WRP evaluation, which included: Julia Larkin, Lucrezia Tincani, Jeremy Colin, Sue Cavill, Sarah Javeed, Alice Mango, Faith Muniale, Shona Jenkins, Alex Hurrell, Sophie Witter, Richard Carter, and Timothy Forster.

REFERENCES

Albertson, K., Fox, C., O'Leary, C , Painter, G., Bailey, K. & Labarbera, J. (2018a) Introduction: outcome-based payment and the reform of public services. In: *Payment by Results and Social Impact Bonds: Outcome-Based Payment Systems in the UK and US*. Policy Press, Bristol. doi:10.1332/policypress/9781447340706.003.0001.

Albertson, K., Fox, C., O'Leary, C., Painter, G., Bailey, K. & Labarbera, J. (2018b) Review of the evidence for outcome-based payment systems. In: *Payment by Results and Social Impact Bonds: Outcome-Based Payment Systems in the UK and US*. Policy Press, Bristol.

Cabinet Office 2011 Open Public Services White Paper. Cabinet Office, London.

Clist, P. 2017 *Review of Payment by Results in DFID: Establishing the Evidence Base*. DFID, London. Available from: https://assets.publishing.service.gov.uk/government/uploads/system/uploads/attachment_data/file/684278/full-report-UEA2-merged.pdf.

DFID 2013 *Water, Sanitation and Hygiene (WASH) Results Programme: Business Case and Intervention Summary*. Available from: https://devtracker.dfid.gov.uk/projects/GB-1-203572/documents.

DFID 2014a *Sharpening Incentives to Perform: DFID's Strategy for Payment by Results*. DFID, London.

DFID 2014b *Designing and Delivering Payment by Results Programmes: A DFID Smart Guide*. Available from: https://assets.publishing.service.gov.uk/government/uploads/system/uploads/attachment_data/file/352519/Designing-Delivering-PbR-Programmes.pdf.

DFID 2016 *Water, Sanitation and Hygiene results programme to support scale-up efforts, Annual Review 2016*. Available from: https://devtracker.dfid.gov.uk/projects/GB-1-203572/documents.

DFID 2018 *Water, Sanitation and Hygiene Results Programme to Support Scale-Up Efforts, Annual Review 2018*. Available from: https://devtracker.dfid.gov.uk/projects/GB-1-203572/documents.

ePact 2018 *WASH Results Programme Evaluation – Synthesis Report Vol. 1.1*. Available from: https://www.gov.uk/dfid-research-outputs/wash-results-programme-evaluation-synthesis-report.

Fox, C. & Morris, S. 2019 Evaluating outcome-based payment programmes: challenges for evidence-based policy. *Journal of Economic Policy Reform*. doi:10.1080/17487870.2019.1575217.

Hood, C. 1991 A public management for all seasons? *Public Administration* **69** (1), 3–19. doi:10.1111/padm.1991.69.issue-1.

National Audit Office 2015 *Outcome-Based Payment Schemes: Government's Use of Payment by Results*. National Audit Office, London.

First received 24 February 2020; accepted in revised form 24 April 2020. Available online 19 May 2020

Research Paper

Benefits and costs of rural sanitation interventions in Ghana

Mark Radin, Brad Wong, Catherine McManus, Saumitra Sinha, Marc Jeuland, Eugene Larbi, Benedict Tuffuor, Noble Kofi Biscoff and Dale Whittington

ABSTRACT

Community-led total sanitation (CLTS) has triggered households around the world to adopt latrines, but evidence suggests that CLTS does not usually lead to universal latrine coverage. Additional interventions, such as subsidies for the poor, may be necessary to eliminate open defecation. While subsidies can improve sanitation-related outcomes, no prior studies have compared the net benefits of CLTS plus subsidies to CLTS-only. This paper presents a comparative analysis for rural Ghana, where efforts to reduce open defecation have had limited success. We analyze the costs and benefits of: (1) a CLTS-only intervention, as implemented in Ghana, and (2) a variant of CLTS that provides vouchers for latrines to the poorest households in high sanitation adoption communities. We find that CLTS-only fails a deterministic benefit-cost test and that only about 30% of 10,000 Monte Carlo trials produce positive net benefits. CLTS plus subsidies satisfy a benefit-cost criterion in the deterministic case, and in about 55% of the Monte Carlo trials. This more favorable outcome stems from high adoption communities passing the threshold needed to generate positive health externalities due to improved community sanitation. The results suggest that a well-targeted CLTS plus subsidies intervention would be more effective in Ghana than CLTS alone.

Key words | community-led total sanitation, benefit-cost analysis, externalities, Ghana, open defecation, WASH

Mark Radin (corresponding author)
Catherine McManus
Department of Environmental Science & Engineering,
Rosenau CB#7341,
Gillings School of Public Health, University of North Carolina at Chapel Hill,
Chapel Hill, NC 27599, USA
E-mail: *maradin@unc.edu*

Brad Wong
Copenhagen Consensus Center,
Tewksbury, MA 01876, USA

Saumitra Sinha
University of North Carolina at Chapel Hill,
Department of City and Regional Planning,
New East Building, CB3140,
Chapel Hill, NC 27599, USA

Marc Jeuland
Sanford School of Public Policy and Duke Global Health Institute,
Duke University,
Box 90239, Durham, NC 27708, USA
and
RWI-Leibniz Institute for Economic Research,
Essen, Germany

Eugene Larbi
Benedict Tuffuor
Noble Kofi Biscoff
TREND,
P. O. Box CT 6135, Cantonments, Accra GA-090-2036, Ghana

Dale Whittington
Department of Environmental Sciences & Engineering, Rosenau CB#7341, Gillings School of Public Health,
University of North Carolina at Chapel Hill,
Chapel Hill, NC 27599, USA
and
Global Research Institute,
University of Manchester,
Manchester M13 9PL, UK

INTRODUCTION

Open defecation (OD) rates have steadily declined across rural Sub-Saharan Africa (SSA) for the past two decades. The percentage of people practicing OD in rural SSA dropped from 43 to 29% from 2000 to 2017 (WHO/UNICEF JMP 2019). During this period, community-led total sanitation (CLTS) gained prominence as one of the

doi: 10.2166/washdev.2020.066

more promising behavior-change interventions for promoting sanitation and was implemented at large or national scale in about 30 countries in the region (Zuin *et al.* 2019). While the introduction of CLTS may have contributed to the decrease in OD in rural SSA, peer-reviewed literature evaluating CLTS interventions in Africa has found that CLTS does not lead to open defecation free (ODF) communities and universal access to latrines (Whittington *et al.* 2020). One potential approach for further reducing OD is to implement variants of CLTS that include additional supportive interventions or incentives. For example, incorporating targeted vouchers or subsidies into CLTS can help reduce OD further, as previously demonstrated in India and Bangladesh (Pattanayak *et al.* 2009; Clasen *et al.* 2014; Patil *et al.* 2014; Guiteras *et al.* 2015; Hammer & Spears 2016).

This benefit-cost analysis (BCA) compares, for the first time, the benefits and costs of implementing a traditional CLTS intervention (CLTS-only) to a variant of CLTS that includes subsidies (CLTS plus subsidies). It builds upon an existing cost-benefit model of CLTS presented in Radin *et al.* (2020). The allows exploration of whether incorporating vouchers increases the net benefits of CLTS plus subsidies, relative to a CLTS-only intervention. We analyze these interventions as if they were implemented in rural Ghana. We choose Ghana as the site of our analysis for several reasons. First, previous work has shown that rates of OD in rural Ghana have been stagnant (decreasing from 32 to 31% between 2000 and 2017), despite increasing development and the country's reclassification as a middle-income country by the World Bank in 2011 (The World Bank 2011; WHO/UNICEF JMP 2019). Second, since Ghana is considered a lower-middle-income country, the cost of implementing CLTS in the country is likely higher than in other countries in SSA, such that the prior analysis presented in Radin *et al.* (2020) seems unlikely to apply (Crocker *et al.* 2017a). Finally, Ghana has continued to invest in efforts to eliminate OD as evidenced by the Social Norms Advocacy Campaign and the National Sanitation Campaign (NSC) to change sanitation norms (UNICEF 2017; Addo 2020), as well as deploying new efforts to target subsidies to the poorest rural households. The NSC includes an ODF Campaign that aims to eliminate OD by 2030 (UNICEF 2017). The ODF Campaign is a component of Ghana's Rural Sanitation Model and Strategy, which endorses CLTS as the main approach for promoting rural sanitation.

Since CLTS was first introduced in Ghana, implementers have observed that many poor and marginalized households struggle to gain access to latrines. In 2018, the Ministry of Sanitation and Water Resources (MSWR) published *Guidelines for Targeting the Poor and Vulnerable* (Guidelines, Ministry of Sanitation and Water Resources 2018) to help address the gap in sanitation access between Ghana's wealthier and poorer households. The Guidelines affirm CLTS as the primary sanitation intervention strategy, but they also suggest subsidizing infrastructure costs to help the poor, and the communities in which they live, to reach universal latrine coverage. The Guidelines specifically call for providing additional support to the poor and vulnerable in communities certified as ODF Basic. The ODF Basic designation is not based on universal coverage with sanitation, but rather means that a community has been verified, by CLTS facilitators, to have no visible fecal matter in public spaces (Ministry of Sanitation and Water Resources 2018). To understand the impact of targeted support to the most vulnerable households, we extend the model from Radin *et al.* (2020) to measure the economic benefits and costs of such a CLTS plus subsidies intervention.

Thus, benefits and costs are estimated and compared for two rural sanitation interventions: (1) the traditional CLTS-only approach and (2) a CLTS plus subsidies variant – inspired by a recent experiment in Ghana (Israel & Shapiro 2019). We specify our analysis to cover a nationally representative set of 100 hypothetical rural communities in Ghana. We model the analysis of the CLTS plus subsidies variation, as it has been implemented thus far in Ghana, i.e., we assume that vouchers are delivered after the CLTS triggering and community mobilization for latrine construction has taken place, and only to communities that show a significant increase in latrine uptake following those activities. Adopting the method used by Radin *et al.* (2020), latrine-uptake rates are assumed to be heterogeneous across communities.

The analysis builds on previous benefit-cost analyses of CLTS and incorporates new information on CLTS implementation in rural Ghana (Whittington *et al.* 2009; Crocker *et al.* 2016a, 2017b; Israel & Shapiro 2019; Radin *et al.* 2020; ODF Ghana Campaign 2017). It serves as an example of how governments and development partners can use context-specific data to evaluate the appropriateness of CLTS, and potential modifications to it, in different locations.

In the second section of this paper, we describe the two sanitation interventions. The third section presents the benefit-cost model. The fourth section discusses some of the key parameter assumptions of the analysis. The fifth section presents the results of the benefit-cost calculations and sensitivity analyses. The sixth and final section concludes.

INTERVENTIONS

Community-led total sanitation

CLTS was first implemented in 1999 and is a community-based intervention that focuses on changing sanitation-related community norms (Kar & Chambers 2008). When it was first developed, CLTS represented a significant policy shift in how to eliminate OD. Instead of simply providing latrines for free or at highly subsidized rates and assuming that they will be used (Jenkins & Sugden 2006), CLTS posits that various community exercises can lead to more sustainable changes in community sanitation behavior and ensure continued demand for latrines. CLTS has been implemented in approximately 60 countries worldwide; about 30 of these countries have incorporated CLTS into their national sanitation strategies (Zuin *et al.* 2019).

CLTS plus subsidies/vouchers

While CLTS remains at the foundation of the rural sanitation program in Ghana, the *Guidelines for Targeting the Poor and Vulnerable* acknowledge that many households have struggled to afford latrines. Numerous CLTS or CLTS-inspired interventions, such as the Total Sanitation Campaign in India, have offered subsidies to the poor in order to boost latrine adoption rates (Pattanayak *et al.* 2009; Clasen *et al.* 2014; Patil *et al.* 2014; Guiteras *et al.* 2015; Hammer & Spears 2016). However, there is no norm or established best practice for the incorporation of subsidies or payment assistance vouchers within CLTS interventions.

The CLTS plus subsidies intervention that we analyze is modeled on an experiment in Ghana as documented by Israel & Shapiro (2019). While the results of the experiment were not available at the time this analysis was conducted, it offers one of the few approaches for ensuring that CLTS can result in universal latrine coverage. In that experiment, vouchers were provided to poor and vulnerable households in villages that had achieved ODF Basic status. For the purposes of this analysis, subsidies are considered to be offered to eligible vulnerable or poor households in select communities within a few months of the CLTS-only triggering in a community (we assume vouchers are distributed between 3 months and 1 year after the CLTS-only intervention based on the time frame of CLTS stages presented in Kar & Chambers (2008) who suggest it takes around 3 months for a community to achieve ODF status). All households who meet eligibility requirements in selected communities receive a voucher to construct a latrine, whether or not they already have one. This affords these low-income households the opportunity to construct a new latrine or to improve or replace an existing one. Importantly, poor households only become eligible for this subsidy support in communities where the CLTS-only intervention achieves high impact.

BENEFIT-COST MODEL AND MODEL ASSUMPTIONS

A BCA is conducted of two sanitation interventions in Ghana over a nationally representative set of 100 rural communities. A detailed description of the equations used to calculate costs and benefits of the two interventions is provided in Supplementary Appendix A, and parameter assumptions are summarized in Supplementary Appendix B.

Common model features for both interventions

We rely on evidence presented in Radin *et al.* (2020) to assume that the response to the CLTS-only intervention is heterogeneous across communities. As in that paper, we assume that there are three levels of uptake (latrine construction) across the 100 communities. In the base case, we assume that 20 of the 100 communities have a large response (high uptake; where latrine coverage increases by 35 percentage points and a community achieves ODF Basic status), 40 communities have a moderate response (medium uptake; where latrine coverage increases by 15 percentage points), and 40 communities have a small response (low uptake; where latrine coverage increases by 5 percentage points) to the intervention.

Accounting for this community-level heterogeneity in uptake affects the calculations of: (1) health impacts of the intervention (externality), (2) time costs of community residents, and (3) latrine-building and operation and maintenance costs. The potential benefits from a positive externality, also known as herd protection, are incorporated in the benefit-cost calculations (for further details, see Radin *et al.* 2020).

The costs and benefits for both sanitation interventions are calculated at the household level in high-, medium-, and low-uptake communities, and then aggregated to the community and regional level. The benefits and costs of the two interventions are calculated for a period of 10 years, and results are reported in terms of two metrics: a benefit-cost ratio (BCR) and a net present value of the time stream of annual benefits and costs over the 10-year planning period. All monetary values are reported in Ghanaian Cedi, GHS (USD 1 = GHS 4.5 in 2018) (Xignite 2018). We also perform three sensitivity analyses: (1) to assess the impact of using different approaches – the value of a statistical life (VSL) or value of a statistical life year (VSLY) methods – to valuing mortality reductions; (2) to assess the impact of using different discount rates, where we follow the guidelines presented in Robinson *et al.* (2019a) and use 8% in the base case (two times the short-term GDP per capita growth of 4%); and (3) to explore an alternative intervention that would provide vouchers to the poorest 40% of households in every community, rather than targeting the poor only in high-uptake communities as is currently done in the experimental intervention we consider in Ghana. We also conduct Monte Carlo simulations to estimate a distribution of likely results, given uncertainty in model parameters. In supplementary appendices to this paper, we present more detailed information about the Monte Carlo simulations (Supplementary Appendix C), a graphical representation of the diarrhea reductions in communities benefiting from the positive sanitation externality (Supplementary Appendix D), and additional tables from the sensitivity analyses (Supplementary Appendix E).

Model features for the CLTS plus subsidies/vouchers variant

For the CLTS plus subsidies intervention, we additionally assume that households induced to build latrines by the voucher receive the same private benefits as households that build latrines without a voucher. The subsidy payment by the government is thus a transfer payment to households receiving the voucher and is not a net cost in the summation of costs and benefits to all stakeholders. This approach overlooks the fact that households that do not build a latrine after the CLTS-only intervention may have lower demand than those that do. In a typical BCA, the value of a latrine ascribed to a household that only obtained it with a voucher would be less than that ascribed to a household that chose to purchase it at full cost; if a household that only obtains a latrine after it receives a voucher valued the latrine as much as a household that obtained a latrine after the CLTS-only intervention or before the intervention, it would have obtained a latrine before the voucher was offered. However, in the context of a rural community in Ghana, households in the bottom two wealth quintiles also face tight liquidity constraints that may make purchasing a latrine unaffordable, even if it is highly desired. In such a context, a lack of access to credit results in suppressed demand.

The voucher scheme affects the total net benefits through four potential pathways: (1) it increases the number of households who build a latrine, such that additional private benefits are derived from these new constructions; (2) in some communities, it may increase coverage past the threshold level at which the health benefits of the positive sanitation externality kick into the benefit of all households; (3) there is an administrative cost to providing the vouchers to the households in the bottom two wealth quintiles in the high-uptake communities where households become eligible for these discounts; and (4) it increases the total capital costs of latrines constructed because more latrines are constructed when vouchers are offered. We include the capital costs of latrines twice for households that newly acquire a latrine and also redeem the voucher because two latrines are constructed for this household. This may overestimate costs somewhat because a poor household that initially built a low-quality private latrine may have incurred lower cost and might be able to upgrade that first latrine at reduced cost, relative to a new upgraded latrine.

PARAMETER ASSUMPTIONS

Assumptions common to both interventions

The model is based around a set of 100 hypothetical average communities in rural Ghana that are targeted by the sanitation program, each having 150 households. These communities have the following characteristics, that are derived from numerous secondary data sources in Ghana: household size (5.37), age composition of a household (0.7 children under 5, 1.5 children between 5 and 14, and 3.2 people 15 and older), daily wages for adults (GHS 4.6 or USD 1.02), diarrheal disease incidence (2.2 for children under 5, 1.15 for children between 5 and 14, and 1.3 for people 15 and older), and case fatality rates for different age groups (0.04% for children under 5, 0.0035% for children between 5 and 14, and 0.02% for people 15 and older) (Crocker *et al.* 2016a, 2017a; Global Burden of Disease Collaborative Network 2017; National Development Planning Commission 2017; Riahi *et al.* 2017; Samir & Lutz 2017; Ghana Statistical Service *et al.* 2018; Radin *et al.* 2020). Therefore, all communities in the analysis are assumed to be observably the same. In reality, communities in rural Ghana are diverse and may vary in both observable and unobservable ways that could affect how communities respond to a CLTS intervention and thus the magnitude of the benefits and costs. Thus, for some districts and communities – and for some households – our estimates of costs and benefits are likely to be too low; for others, they will be high.

Our analysis is based on national data describing conditions in rural Ghana. However, CLTS has already been implemented in every region (but not every community) in Ghana. There may be fundamental differences between communities where CLTS has been implemented and those where it has never been implemented. As such, though the characteristics that we specify to construct each model community are representative of rural Ghana, they may not be representative of the rural communities where CLTS has not previously been implemented.

Baseline latrine coverage

Data from the WHO/UNICEF Joint Monitoring Program (2020) are used to estimate baseline private latrine coverage by wealth quintile. The benefit-cost calculations assume that the CLTS-only intervention has the same effect on poor and non-poor households, i.e., the same proportion of households from each wealth quintile constructs latrines as a result of the program. We assume the impact is proportional because evidence on how wealth status mediates the impact of CLTS is limited. Cameron *et al.* (2019) find a greater impact among non-poor households, while Pickering *et al.* (2015) finds a greater impact among the poor. We, therefore, assume that 60% of the latrine increase occurs in the households in the top three wealth quintiles (60% of the population), while the other 40% occurs in households in the bottom two wealth quintiles (40% of the population). We also assume that an equal proportion of those practicing OD and using shared latrines at baseline shift to using a private single-household latrine following the intervention. At baseline, approximately 30% of the population engages in OD and 35% uses shared latrines. Of the total population without a private single-household latrine, 46% practice OD (30/65) and 54% (35/65) use a shared latrine. Therefore, 46% of the households adopting a private single-household latrine are assumed to switch from OD, while 54% are households switching from using shared sanitation. There is ongoing debate about the health benefits of shared latrines (Heijnen *et al.* 2014); however, for the purposes of this analysis, it is assumed that shared latrines confer no protection from diarrhea.

Program costs

The program costs for the CLTS program are assumed to be independent of the level of latrine uptake as most of the intervention costs are spent delivering the intervention, before households have had time to respond to the intervention. However, the time costs for participating in the initial triggering and follow-up activities vary with uptake. Not all households in the communities are assumed to attend triggering activities. Communities where more households attend the meetings have a higher time cost than those that have lower attendance rates. The cost per latrine is assumed to be constant, but the total community costs of building latrines – and the community-level operation and maintenance costs – are higher in communities that have more uptake. Based on empirical evidence from field evaluations in other countries, we assume that for the first 5 years after the intervention, all households constructing a latrine will use their latrine. After 5 years, a constant rate of households that have built private latrines in response to the intervention are assumed to abandon them (Barnard *et al.* 2013; Crocker *et al.* 2017b; Cameron *et al.* 2019; Orgill-Meyer *et al.* 2019), which leads to a reduction in benefits as well as a decline in operation and maintenance costs over time. Program costs (including management, training, and facilitation) and local investments (including time and money related to community activities and latrine constructions) are based on estimates by Crocker *et al.* (2017a).

Economic value of health benefits and costs

This paper follows the recommendations for health valuation presented in Robinson *et al.* (2019a). We use VSL to estimate the benefits of averted mortality from diarrhea cases. We use a benefit transfer approach to estimate the VSL assuming: (1) GNI per capita (PPP) in the United States (USD 61,019); (2) GNI per capita in purchasing power parity of Ghana in 2017 International Dollars 4,864; (3) an income elasticity for VSL of 1.5; (4) a U.S. VSL of 2017 International Dollars 9.6 million; (5) a growth rate in Ghana that ranges from 4.1 to 3.6% per year over the life of the project, and (6) a 2017 GHS to International Dollar conversion factor. These assumptions imply a VSL of GHS 413,220 (USD 91,827) in 2018 ($413,220_{VSL\ Ghana} = 9,600,000_{VSL\ USA} * (((4,864_{Ghana\ GNI\ per\ capita\ 2017}) / (61,019_{USA\ GNI\ per\ caita\ 2017})) * 1.041_{\text{Ghana real growth}})^{1.5} * 1.8_{\text{PPP Conversion Factor Ghana 2017}})$.

In a sensitivity analysis, we compare the results of this VSL approach to those obtained from a VSLY approach. The latter specifies a constant value for each additional year of life gained due to the intervention, which is calculated based on the VSL – described above – and the average life expectancy in the population of interest (about 65 years for children under 5, 60 years for children between 5 and 14, and 41 years for people 15 and older). The VSLY is estimated to equal GHS 12,556 (USD 2,790). The results of an analysis using VSLY and VSL are likely to differ when the population most affected by an intervention is either much older or much younger than the average person. The benefits will be higher using the VSLY approach if a large number of younger people are affected by an intervention and lower if a small number of older people are affected because preventing the death of a younger person results in more additional years lived in expectation (Robinson *et al.* 2019b). Since diarrhea morbidity and mortality are higher for children under 5, we expect and find that the VSLY analysis has a more favorable benefit-cost result compared with the VSL analysis.

The cost of illness (COI) method is used to value the averted non-fatal morbidity from diarrhea cases, as no willingness-to-pay measure is available (Robinson & Hammitt 2018). The COI includes all direct and indirect costs of getting sick with diarrhea, including the costs paid by the healthcare system and lost productivity time. Based on Nonvignon *et al.* (2018) and the 2017 Maternal Health Survey (Ghana Statistical Service *et al.* 2018), we assume COI for an individual seeking care to be GHS 6.70 (USD 1.48) for outpatient treatment and GHS 87.92 (USD 19.42) for inpatient treatment. Similarly, the cost of a single case of diarrhea to the healthcare system is GHS 15.02 (USD 3.32) for outpatient treatment and GHS 195.89 (USD 43.27) for inpatient treatment. The overall weighted COI estimates per case of diarrhea based on the proportion of people seeking care and inpatient and outpatient cases disaggregated by age are GHS 86 (USD 19) for children under 5, GHS 14 (USD 3) for children 5–14, and GHS 26 (USD 6) for people 15 and older.

Finally, we follow the recommendation of Whittington & Cook (2019) to estimate the value of time as 50% of the local informal wage rate, approximately GHS 40 (USD 8.84) per day, for adults. We assume a value of time for children 5–14 of 25% of the local informal wage rate and do not value the time savings of children under 5. We assume that individuals save 5 min/day when switching from OD to a private single-household latrine or from a shared latrine to a private single-household latrine (J. Crocker, personal communication, November 6, 2018).

Community response to CLTS activities

While many factors may influence the community response to CLTS, the model assumes that the only difference between low-, medium-, and high-uptake communities is their response to the CLTS-only intervention. The proportion of each of these types of communities in the hypothetical region is based on evidence discussed in Radin *et al.* (2020). The empirical evidence on the heterogeneity of community-level responses does not provide any insight into other, systematic differences between these types of communities. In other words, we do not assume that high-uptake communities, for example, have different rates of diarrheal disease prevalence than low-uptake communities.

Effects of the CLTS intervention

Radin *et al.* (2020) reviewed 14 studies on CLTS and CLTS-inspired interventions and documented increases in latrine construction ranging from 0 to 50 percentage points, with an average increase of approximately 15 percentage points (Pattanayak *et al.* 2009; Elbers *et al.* 2012; Cameron *et al.* 2013; Clasen *et al.* 2014; Patil *et al.* 2014; Guiteras *et al.* 2015; Pickering *et al.* 2015; Crocker *et al.* 2016a, 2016b; Hammer & Spears 2016; Briceño *et al.* 2017; Luby *et al.* 2018; Null *et al.* 2018; Humphrey *et al.* 2019). In this analysis, we assume an average increase of private household latrine coverage across the full set of communities of 15 percentage points: 5 percentage points for the 40% of communities with low uptake, 15 percentage points for the 40% of communities with medium uptake, and 35 percentage points for the 20% of communities with high uptake.

We consider unimproved latrines to be private single-household latrines, but shared latrines do not count as private single-household latrines. The adoption of an unimproved or an improved private single-household latrine is assumed to confer health benefits to all people within a household. While many CLTS studies have found that CLTS interventions have a limited impact on diarrhea, some studies have found statistically significant effects (Hammer & Spears 2016; Cameron *et al.* 2019). We adopt the approach used in Radin *et al.* (2020) and assume an average reduction of 20% of the baseline diarrhea rate.

Positive sanitation externality

Evidence for the presence of a sanitation externality has been discussed in a number of studies in the literature on the health impacts of sanitation interventions. These studies attempt to establish the link between higher rates of community-level improved sanitation behavior and greater reductions in diarrheal disease (Andres *et al.* 2017; Harris *et al.* 2017; Jung *et al.* 2017; Wolf *et al.* 2018). While this herd protection may have benefits that extend beyond reducing diarrheal disease incidence, due to limited data we only consider reductions in diarrhea (Fuller *et al.* 2016). We follow the assumptions presented in Andres *et al.* (2017) and Wolf *et al.* (2018), as well as Radin *et al.* (2020) who operationalize this concept for BCA, specifying a threshold level of 75% coverage beyond which the externality is realized. In the communities where this 75% threshold is exceeded, we estimate the diarrheal risk reduction as follows. Households that construct a new latrine due to the intervention experience a diarrheal risk reduction that increases linearly from 20%, when there are no externality benefits (i.e., when the latrine coverage level in the community is below the 75% threshold), to a maximum of 35% when community coverage reaches 100%. Households in communities with private single-household latrine coverage exceeding 75%, but that do not adopt latrines themselves, experience a diarrheal risk reduction that increases linearly from 0 to 35% as a function of community latrine coverage. Finally, households that already owned private single-household latrines prior to the intervention are assumed to receive a more limited benefit from the externality – a risk reduction rising linearly from 0 to 20% above the threshold – as they previously already experienced the private health benefits of adopting a private single-household latrine. We assume that households that already had a latrine before the CLTS-only intervention but choose to redeem the voucher to obtain a new latrine do not experience any additional health benefits beyond the health externalities that arise once community-level latrine coverage passes the 75% threshold. We present a graphical description of the diarrhea reductions in Supplementary Appendix D. Furthermore, we note here that due to the low baseline sanitation coverage in rural Ghana, this externality is only realized in the second intervention (CLTS plus subsidies) because the assumed impact of the CLTS-only intervention, even in high-uptake communities, does not meet the required externality threshold.

Assumptions specific to the CLTS plus subsidies/vouchers intervention

The literature on the effect of including subsidies within a CLTS intervention is limited. While a number of studies document the potential for subsidies to complement CLTS, there are no existing benefit-cost studies that demonstrate the additive effects of a post-CLTS voucher program (Pattanayak *et al.* 2009; Patil *et al.* 2014; Guiteras *et al.* 2015).

Importantly, we do not evaluate the appropriateness of the design or implementation of the ongoing subsidy program (Israel & Shapiro 2019). Rather, the benefits and costs associated with the intervention as it is designed are calculated, assuming that the targeting approach implemented is successful and does not lead to perverse incentives or other outcomes. However, we also comment on potential challenges with this targeting policy in the discussion.

Subsidy eligibility

One important difference in the CLTS plus subsidies intervention included in this analysis and subsidy interventions that have been tried elsewhere is that the subsidies are provided to poor households only *after* the CLTS-only intervention. Because the subsidy disbursement takes place after the CLTS-only intervention, and only if a community has been sufficiently responsive, a relatively large number of households will already have built latrines prior to the poorer community members becoming eligible for subsidies.

Consistent with the policy experiment on which our intervention is based, the CLTS plus subsidies intervention is assumed to provide vouchers to a fraction of households only in those communities where the CLTS-only intervention has already achieved a high impact. Following Radin *et al.* (2020), it is assumed that among the 100 model communities, there are three types of responses: 20 communities have a large response (high uptake; where latrine coverage increases by 35 percentage points and a community achieves ODF Basic status), 40 communities have a moderate response (medium uptake), and 40 communities have a small response (low uptake) to the intervention.

The subsidy component of the CLTS plus subsidies intervention then only applies in the high-uptake communities. In these high-uptake communities, the subsidy provides vouchers that cover 100% of the cost of improved latrines for the poorest 40% of households, regardless of whether or not they have already constructed a latrine. In practice, household eligibility rates in different communities would perhaps be determined based on household wealth, which could result in variation in the share of the subsidy-eligible population across high-uptake communities. Absolute measures of poverty status could be estimated using information on eligibility for existing government programs that are implemented throughout Ghana, such as the National Health Insurance Scheme indigent exemption waiver or Livelihood Empowerment Against Poverty (LEAP) program, while relative measures (to identify the poorest 40% within any given high-uptake community) could be based on guidance from local community leaders and the CLTS committee formed as part of the CLTS-only intervention (Palermo *et al.* 2019). Vouchers, equivalent to the cost of materials for one latrine, could be redeemed at local shops or retailers who would then collect payment from the CLTS implementers and provide receipts of sales to verify latrine purchases. This protocol generally adheres to the Guidelines established by the MSWR, which states that assistance (in this case, subsidies or vouchers) is only to be delivered to poor households in communities which can be classified as high-uptake.

Voucher redemption

The Guidelines include minimum standards for latrines and identify appropriate technology options, including ventilated improved pit latrines and pour flush latrines (Ministry of Sanitation and Water Resources 2018). Past studies of CLTS have shown that newly built latrines in intervention communities may use low-quality materials and suffer from durability problems (Cavill *et al.* 2015; Crocker *et al.* 2017b). Thus, poor households that previously had or built new latrines during the CLTS-only intervention might still choose to redeem the voucher to obtain newer and higher-quality latrines. We have limited information on who would redeem these vouchers, but existing evidence suggests it would not be everyone, due to individual preferences, supply constraints, or difficulties in redeeming vouchers (Banful 2011; Wright *et al.* 2016). Lacking Ghana-specific data, we draw on evidence from voucher experiments in rural Kenya and Tanzania (Peletz *et al.* 2017, 2019) to account for the incomplete adoption of subsidies and latrines. Specifically, it is assumed that in high-uptake communities, 80% of subsidy-eligible households (the 40% in the bottom two wealth quintiles) redeem the voucher. This 80% includes those who owned the latrine before the CLTS-only intervention, those who built a new latrine because of the CLTS-only intervention, and those who still practice OD after the CLTS-only intervention. Therefore, households who redeem the voucher do so to either move from practicing OD to using a latrine or from using a lower-quality latrine to a higher-quality latrine. Households that have a newly built a latrine due to the CLTS-only intervention that redeem the vouchers can upgrade their latrine and obtain a higher-quality latrine that adheres to the Guideline standards. Thus, total coverage with improved sanitation among the bottom two wealth quintiles would reach 80%. We allow this redemption rate to range from 70 to 90% in the sensitivity analysis. This redemption rate considers that some poor households build high-quality latrines after the CLTS-only intervention and so do not redeem the voucher.

We assume that the 80% of households in the bottom two wealth quintiles who redeem the vouchers come proportionately from three types of households: (1) those who have access to a private single-household latrine, (2) those who have access to a shared latrine, and (3) those who practice OD. Importantly, however, we assume that the proportions of voucher-redeeming households in each of these three groups are somewhat different than for the community as a whole, reflecting the lower improved sanitation coverage among those with low incomes. Specifically, we assume that at baseline, half of the households in the bottom two wealth quintiles practice OD (20% of the total population). After the CLTS component of the CLTS plus subsidies intervention in the high-uptake communities and before the subsidy component of the intervention is implemented, we assume that 30% of the households in the bottom two wealth quintiles practice OD (or 12.5% of the total population). Once subsidies are disbursed, 80% of this 30% of households in the bottom two wealth quintiles switch to using an improved latrine while the rest continue to practice OD (2.5% of the total population). Although Guiteras *et al.* (2015) report that subsidies to poorer houses will also result in behavior changes in the wealthier households, we do not assume any additional change among non-poor households due to the implementation of the voucher program.

To summarize, this second intervention includes the entire CLTS-only intervention plus a post-CLTS provision of vouchers to the 40% poorest households in the 20 high-uptake communities in the hypothetical region.

Program costs and benefits

The redeemed vouchers are assumed to shift the cost of latrine construction from the household onto the funder of the CLTS plus subsidies program since the retailers from whom vouchers are exchanged receive payment from funders. A fixed administrative cost is incurred by implementers in each community where the subsidy vouchers are provided. This is the only change to program costs.

In the low- and medium-uptake communities, benefits accrue only to those households that build latrines. In high-uptake communities, however, benefits accrue to all households because community latrine coverage surpasses the threshold that is assumed to be necessary to experience the positive sanitation externality.

RESULTS

For both the CLTS-only and the CLTS plus subsidies interventions, Table 1 shows the number of non-fatal statistical cases of diarrhea avoided, premature deaths averted, and hours saved by each age group for the low-uptake, medium-uptake, and high-uptake communities and at the

Table 1 | Estimates of cases of diarrhea avoided, premature deaths averted, and hours saved – for CLTS-only and CLTS plus subsidies interventions (totals over 10-year planning horizon)

	Low-uptake community	Medium-uptake community	High-uptake community	All communities ($n = 100$)
CLTS-only				
Statistical cases avoided total[a]	103	309	721	30,910
<5	21	64	150	6,430
5–14	24	73	169	7,240
≥15	58	172	402	17,240
Premature deaths averted total	0.021	0.063	0.15	6.3
<5	0.009	0.026	0.060	2.6
5–14	0.001	0.002	0.006	0.2
≥15	0.012	0.035	0.081	3.5
Hours saved total	265	796	1,856	79,550
5–14	85	256	597	25,600
≥15	180	540	1,259	53,950
CLTS plus subsidies				
Statistical cases avoided total[a]	103	309	1,179	40,065
<5	21	64	245	8,335
5–14	24	73	276	9,380
≥15	58	172	658	22,350
Premature deaths averted total	0.021	0.063	0.24	8.1
<5	0.009	0.026	0.10	3.3
5–14	0.001	0.002	0.0089	0.30
≥15	0.012	0.035	0.13	4.5
Hours saved total	265	796	2,440	91,230
5–14	85	256	785	29,360
≥15	180	540	1,655	61,870

[a]We note here that the statistical cases avoided refers to non-fatal diarrhea cases.

regional level over the 10-year planning horizon. We estimate that in the region with 100 communities, the CLTS-only intervention leads to a decrease of 30,910 non-fatal statistical diarrhea cases, averts a total of 6 premature deaths, and saves 79,550 h over the 10-year expected life of the latrines built due to the intervention. When subsidies are offered, we estimate that in the region of 100 communities, the intervention leads to a decrease of about 40,065 statistical diarrhea cases, a total of 8 premature deaths are averted, and 91,230 h are saved over the 10-year expected life of the latrine.

Table 2 presents the benefits, costs, NPV, BCR, and NPV per household targeted, and per household adopting a latrine for the two interventions for communities with low, medium, and high uptake of private single-household latrines calculated at a discount rate of 8%. Across the full set of 100 hypothetical Ghanaian communities, the present value of the total benefits of the CLTS-only intervention is GHS 3,643,915 (USD 809,759), an amount that increases to GHS 5,212,990 (USD 1,158,442) when subsidies are offered. The present value of the costs of the CLTS-only intervention in the region of 100 communities is GHS 3,865,050 (USD 858,900), which increases to GHS 4,681,175 (USD 1,040,261) when subsidies are offered. This cost increase is due to the cost of administering the subsidy program in eligible communities, the costs of providing subsidies to households in the bottom two wealth quintiles, and the cost of the additional latrines built for households newly adopting latrines and for households upgrading latrines. The BCR of the CLTS-only intervention in the region is 0.94 and increases to 1.1 for CLTS plus subsidies. The overall net benefits per household in the region (the

Table 2 | Summary of results of BCA for CLTS-only and CLTS plus subsidies interventions: low, medium, and high-uptake communities (8% discount rate)

	Low-uptake community	Medium-uptake community	High-uptake community	All communities ($n = 100$)
CLTS-only				
Benefits	GHS 12,145	GHS 36,440	GHS 85,025	GHS 3,643,915
Mortality benefits	GHS 8,635	GHS 25,905	GHS 60,445	GHS 2,590,540
Morbidity benefits	GHS 3,105	GHS 9,315	GHS 21,730	GHS 931,380
Time savings	GHS 405	GHS 1,220	GHS 2,850	GHS 121,995
Costs	GHS 23,835	GHS 38,905	GHS 67,780	GHS 3,865,050
Program costs	GHS 15,625	GHS 15,625	GHS 15,625	GHS 1,562,445
Time costs	GHS 1,330	GHS 2,635	GHS 3,985	GHS 238,305
Capital costs	GHS 4,115	GHS 12,355	GHS 28,825	GHS 1,235,250
O&M costs	GHS 2,765	GHS 8,290	GHS 19,345	GHS 829,050
Net benefits	(GHS −11,690)	(GHS −2,465)	GHS 17,245	(GHS −221,135)
BCR	0.51	0.94	1.3	0.94
Net benefits per HH	(GHS −80)	(GHS −16.5)	GHS 115	(GHS −15)
Net benefits per latrine adopting HH	(GHS −1,560)	(GHS −110)	GHS 330	(GHS −100)
CLTS plus subsidies				
Benefits	GHS 12,145	GHS 36,440	GHS 163,475	GHS 5,212,990
Mortality benefits	GHS 8,635	GHS 25,905	GHS 99,650	GHS 3,374,655
Morbidity benefits	GHS 3,105	GHS 9,315	GHS 36,085	GHS 1,218,435
Time savings	GHS 405	GHS 1,220	GHS 3,740	GHS 139,900
Voucher	GHS 0	GHS 0	GHS 24,000	GHS 480,000
Costs	GHS 23,835	GHS 38,905	GHS 108,585	GHS 4,681,175
Program costs	GHS 15,625	GHS 15,625	GHS 15,625	GHS 1,562,445
Time costs	GHS 1,330	GHS 2,635	GHS 3,985	GHS 238,305
Capital costs	GHS 4,115	GHS 12,355	GHS 38,020	GHS 1,419,160
O&M costs	GHS 2,765	GHS 8,290	GHS 25,515	GHS 952,485
Voucher	GHS 0	GHS 0	GHS 24,000	GHS 480,000
Voucher administration	GHS 0	GHS 0	GHS 1,440	GHS 28,780
Net benefits	(GHS −11,690)	(GHS −2,465)	GHS 54,890	GHS 531,815
BCR	0.51	0.94	1.5	1.1
Net benefits per HH	(GHS −80)	(GHS −16.5)	GHS 365	GHS 35.5
Net benefits per latrine adopting HH	(GHS −1,560)	(GHS −110)	GHS 1,045	GHS 235

total net benefits of the interventions divided by the total number of households, which is 15,000 households) was GHS −15 (USD −3.3) for the CLTS-only intervention and GHS 35.5 (USD 7.9) for the CLTS plus subsidies intervention. The net benefits per household adopting latrines (the total net benefits of the interventions divided by the total number of households who built latrines, which is 2,250 households) was GHS −100 (USD −22) for the CLTS-only intervention and GHS 235 (USD 52) for the CLTS plus subsidies intervention.

As detailed in Table 2, there is a large difference in benefits by community type. In communities with high uptake, the CLTS-only intervention has a BCR of 1.3 and the CLTS plus subsidies intervention has a BCR of 1.5. In medium-uptake communities, the BCR is the same (0.94) for both the CLTS-only and the CLTS plus subsidies intervention because no subsidies are provided. Similarly, in low-uptake communities, the BCR for both interventions is 0.51. While increasing the proportion of high-uptake communities would make the interventions much more beneficial, it is very difficult to identify these communities before implementation.

We present the results of our sensitivity analyses using different discount rates and VSLY to measure mortality benefits rather than VSL as well as a more detailed distribution of the costs and benefits to each of the stakeholders in Supplementary Appendix E.

Table 3 presents the results of the analysis assuming all households in the bottom two wealth quintiles in all communities within the region receive vouchers. This analysis shows slightly higher NPVs and BCRs than the results when subsidies are only targeted to the high-uptake communities. These results are slightly more favorable because the subsidies lead to large increases in uptake in the low-uptake and medium-uptake communities (for low-uptake communities latrine coverage increases from 39 to 60% and for medium-uptake communities from 49 and 66%), and many more households benefit from improved sanitation. The increase in uptake additionally brings medium-uptake communities very close to the externality threshold. In reality, this threshold is uncertain and whether it would be induced by wider targeting of pro-poor subsidies is an important question. Of course, these results assume that voucher redemption rates within a community are independent of the size of latrine uptake due to the intervention, which may not be the case.

The results of the Monte Carlo analysis show that a wide range of outcomes are possible for the two CLTS interventions. Figure 1 presents the cumulative density functions of the NPVs and BCRs, for the two interventions (CLTS-only and CLTS plus subsidies). The Monte Carlo simulation shows that the NPV of the CLTS-only intervention is greater than zero in fewer than 30% of simulations. The NPV of CLTS plus subsidies is greater than zero in about 55% of simulations. These results suggest that the Government of Ghana is likely to have other investment opportunities that are more attractive than both of these CLTS interventions, though CLTS with subsidies appears more attractive than CLTS alone.

We also present a one-way sensitivity analysis, where we only vary the value of one parameter while keeping all others constant and identify the ten parameters that explain the most variation in these NPV results (Figure 2).

Table 3 | Summary of results of BCA with subsidy/voucher for all: low, medium, and high-uptake communities (8% discount rate)

	Low-uptake community	**Medium-uptake community**	**High-uptake community**	**All communities ($n = 100$)**
CLTS plus subsidy				
Benefits	GHS 85,370	GHS 101,660	GHS 163,475	GHS 10,750,970
Mortality benefits	GHS 44,230	GHS 55,970	GHS 99,650	GHS 6,000,980
Morbidity benefits	GHS 15,900	GHS 20,120	GHS 36,085	GHS 2,162,680
Time savings	GHS 1,240	GHS 1,570	GHS 3,740	GHS 187,310
Voucher	GHS 24,000	GHS 24,000	GHS 24,000	GHS 2,400,000
Costs	GHS 77,755	GHS 88,450	GHS 108,590	GHS 8,819,845
Program costs	GHS 15,625	GHS 15,625	GHS 15,625	GHS 1,562,445
Time costs	GHS 1,330	GHS 2,635	GHS 3,990	GHS 238,300
Capital costs	GHS 21,160	GHS 26,780	GHS 38,020	GHS 2,677,895
O&M costs	GHS 14,200	GHS 17,970	GHS 25,515	GHS 1,797,295
Voucher	GHS 24,000	GHS 24,000	GHS 24,000	GHS 2,400,000
Voucher administration	GHS 1,440	GHS 1,440	GHS 1,440	GHS 143,910
Net benefits	GHS 7,615	GHS 13,210	GHS 54,885	GHS 1,931,125
BC ratio	1.1	1.1	1.5	1.2

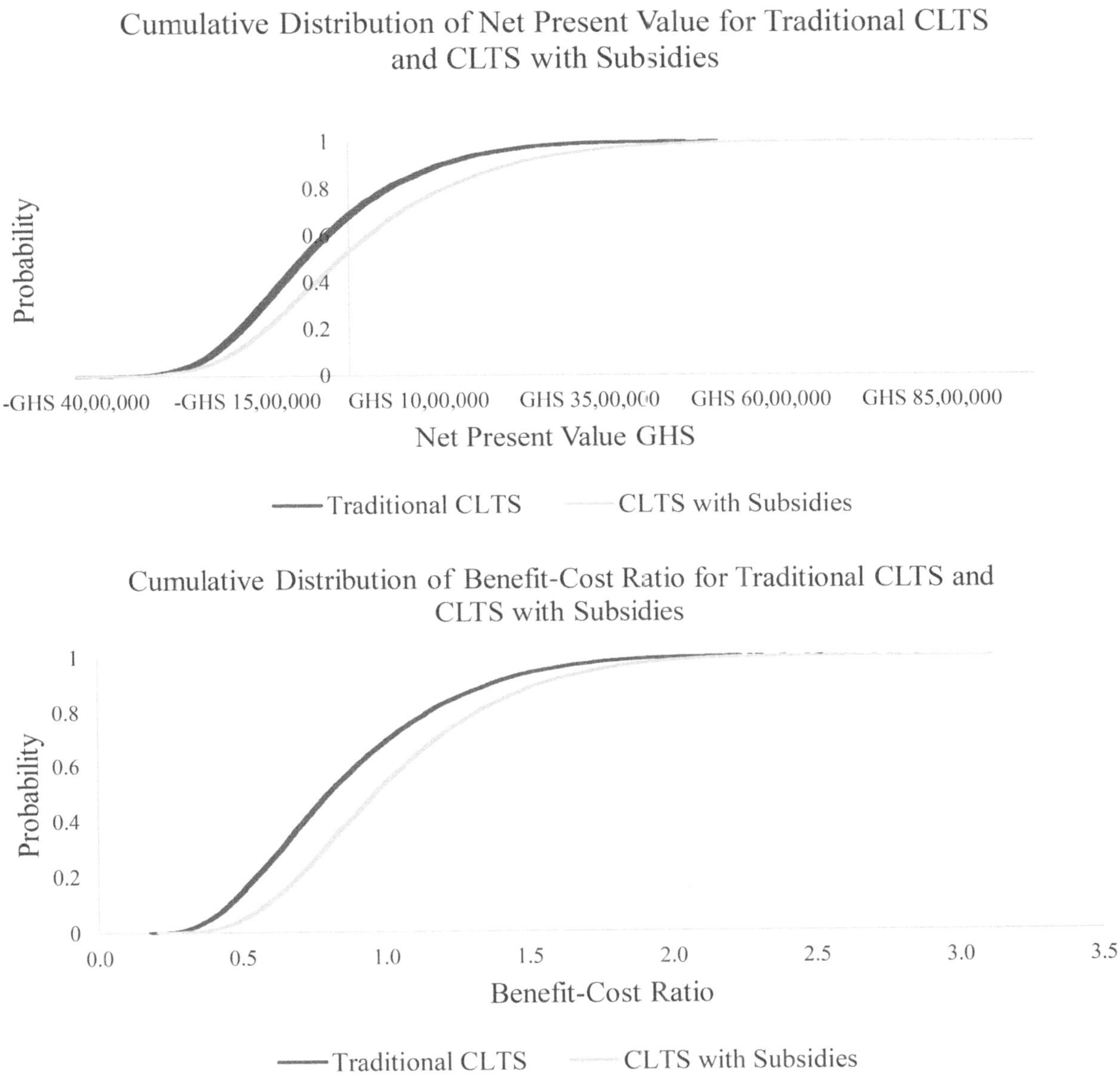

Figure 1 | Cumulative distribution of results from Monte Carlo simulation (10,000 draws) of traditional CLTS and CLTS with subsidies: NPV and BCR.

For both interventions, the diarrhea reduction for households newly adopting latrines, the case fatality rate, and the cost of the latrine were the three most influential variables, in that order. For CLTS-only, the next most influential variables are (1) the discount rate, (2) the cost of operation and maintenance as a percent of the total latrine cost, (3) CLTS facilitation costs, (4) baseline diarrheal incidence rate, (5) the percent of communities with low-uptake, (6) the percentage point increase in latrines in low-uptake communities, and (7) the percentage point increase in latrines in medium-uptake communities. For CLTS with subsidies, the next most influential variables are (1) the percent of communities with low uptake, (2) the percentage point increase in latrines in high-uptake communities, (3) the externality threshold, (4) the discount rate, (5) the cost of operation and maintenance as a percentage of the total latrine cost, (6) the baseline diarrheal incidence, and (7) CLTS

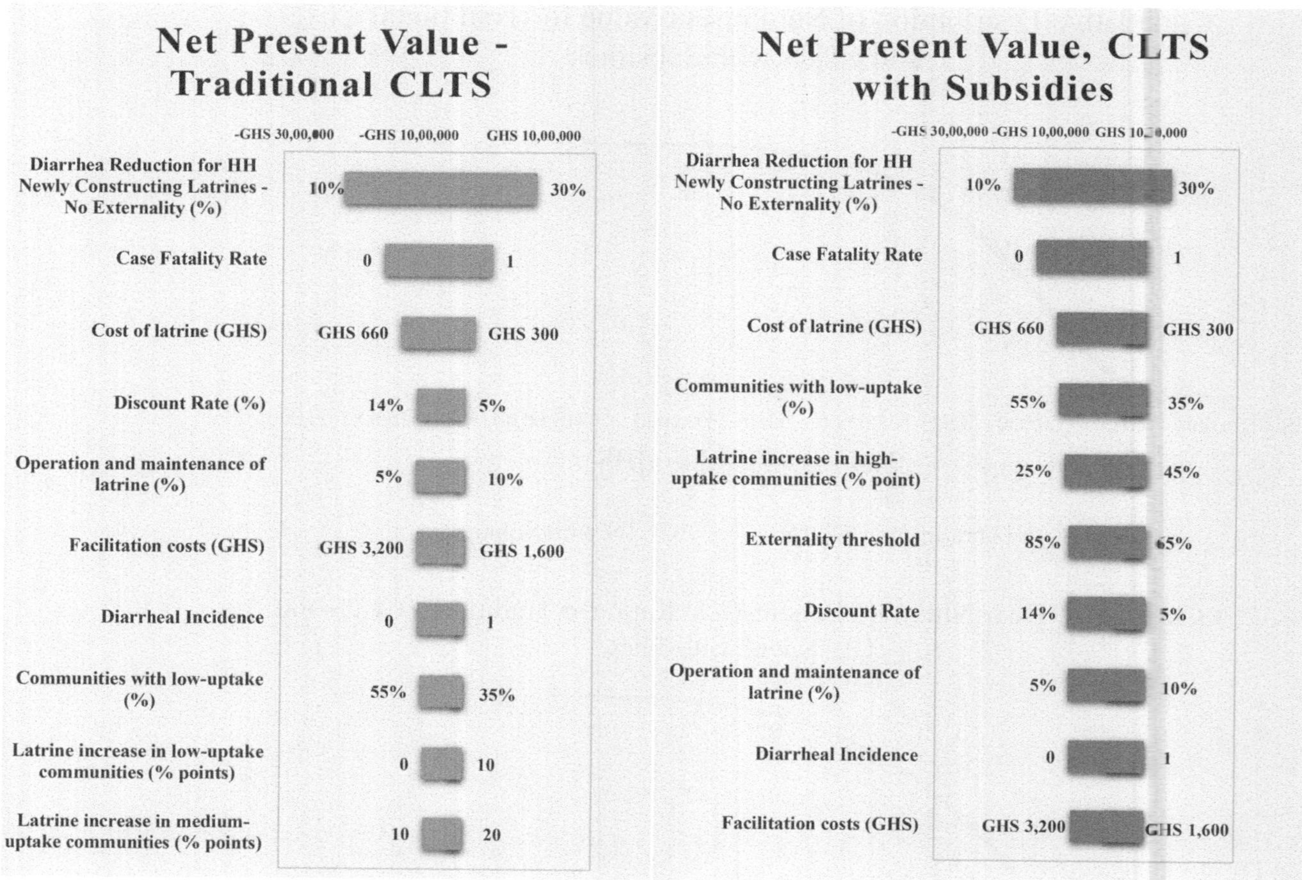

Figure 2 | Sensitivity analyses: effect of selected parameters on NPV with and without externality (holding other parameters at base case values).

facilitation costs. Such sensitivity analyses may help CLTS program implementors identify those communities where net benefits are most like to be high. For example, communities with higher diarrheal morbidity and mortality, or where latrines implementing CLTS would be cheaper, or where the response to the CLTS intervention would be greatest are good candidates to receive a CLTS intervention. However, to use such guidance in practice to target the activities of a future CLTS campaign may be difficult as more remote communities are likely to have higher diarrheal morbidity and mortality while less remote communities are likely to have cheaper latrine and implementation costs. Furthermore, there is insufficient evidence to allow identification of those communities that would have the largest response to a CLTS intervention.

DISCUSSION

When deciding whether to invest in a new CLTS intervention, planners should consider the local realities in their specific program area, which may be more or less favorable for the intervention than the conditions analyzed here. And so, the results of the benefit-cost calculations presented here should be viewed as potential outcomes, given the substantial heterogeneity in baseline sanitation infrastructure, health, and socioeconomic conditions across rural communities in Ghana. Moreover, since CLTS has already been implemented in every region in Ghana, it is possible that the government officials and development organizations strategically chose to first implement CLTS in communities where they believed it would be most effective. An analysis for a specific region of Ghana or a different country, particularly one in which costs would be lower than in Ghana, might yield different results (Crocker *et al.* 2017a; Cha *et al.* 2020).

This analysis suggests that communities with average diarrheal morbidity and mortality rates and similar costs to those assumed here will likely not experience large positive net benefits from the CLTS-only intervention. For Ghana, implementing CLTS-only interventions in many communities would not achieve the goals of ending OD set forth by the NSC. Experimenting with complementary interventions, however, such as the inclusion of a targeted subsidy provision, could lead to the development of a package of CLTS and subsidies that would achieve the NSC goals.

However, the evidence base on the relative effectiveness of various subsidy-related modifications to the CLTS program is thin. Little is known about the relative merits of different types of sanitation subsidy programs and how the following features affect outcomes: (1) selection criteria for identifying eligible beneficiaries, (2) type and amount of subsidy, (3) time point within the intervention at which the support is provided, and (4) how the subsidy is disbursed (Pattanayak *et al.* 2009; SNV 2010; Boisson *et al.* 2014; Venkataramanan 2014; Guiteras *et al.* 2015; Coffey *et al.* 2017; Garn *et al.* 2017; Augsburg *et al.* 2019; Batmunkh *et al.* 2019). Moreover, the broader literature in economics on the challenge of delivering subsidies (see, e.g., Devereux *et al.* 2017) points to potential problems related to the type of targeting being implemented in Ghana.

As discussed, our analysis of the CLTS with subsidies intervention is based on an experiment currently being implemented in rural Ghana. That experiment offers vouchers only to poor households in communities that have been declared ODF Basic (Ministry of Sanitation and Water Resources 2018; Israel & Shapiro 2019). This might appear odd at first glance – why offer subsidies where universal sanitation has already been achieved without them? However, in practice, the ODF declaration does not require universal coverage with high quality (improved) sanitation in Ghana. Nonetheless, this particular policy has several potential drawbacks. First, as word spreads about the design of the incentive, households that are eligible to receive the subsidy will have an incentive to hold out until the subsidy is administered, which would suppress the gains from CLTS efforts. This also means that poor households who do construct latrines and contribute to reaching a minimum coverage needed to receive the subsidies may construct lower-quality ones because of lack of initial funding, which would be wasteful spending if these are then replaced when subsidies arrive.

Second, there is also an important equity concern. Poor households that are effectively 'triggered' and spend significant resources on latrines may resent a policy that delivers subsidies to their less involved neighbors. Third, communities may engage in strategic behavior, organizing in ways that superficially meet the 'high uptake' standards for receiving vouchers without making real behavior change, which is what is needed for a CLTS intervention to truly reduce health risks. Fourth, when scaling-up the policy to the national level, administrative hurdles for this type of nuanced policy, which requires effective monitoring and timely disbursement of subsidies to maintain momentum toward ending OD, are likely to be high. The broader evidence in support of conditional subsidies, rather than unconditional ones, is mixed, and details of design and implementation matter a great deal (SNV 2010; Venkataramanan 2014; Guiteras *et al.* 2015; Coffey *et al.* 2017; Garn *et al.* 2017; Augsburg *et al.* 2019).

BCA estimates the social benefits and costs, not just costs to program implementers. The total costs presented in this analysis are not all borne by implementers. This distinction is important for at least two reasons. First, the government and implementers must develop budgets for CLTS programs that are feasible from a public sector or donor budgeting perspective. The CLTS intervention that includes subsidies costs an additional GHS 480,000 (USD 106,031) that the government or other implementers would somehow need to pay for. However, from the perspective of a social BCA, these costs are not additional as they only shift the subsidized costs from the household ledger to that of the implementer. Therefore, assuming that the subsidy is efficient and well-targeted, reducing the size or number of households receiving a subsidy only affects the overall benefit-cost results through reductions in uptake (and therefore the extent to which the positive externality is captured), but it would affect the implementers' budget. There are also challenges with implementing subsidy schemes to limit leakage, which we have are not incorporated into this analysis.

Second, private costs borne by households – in terms of time spent participating in CLTS activities, money or in-kind expenditures for construction of latrines, and even aesthetic costs associated with using lower quality on-site sanitation facilities – are precisely the types of barriers that prevent more widespread adoption of sanitation improvements (Guiteras *et al.* 2015; Coffey *et al.* 2017; Crocker *et al.* 2017a). Subsidies can play an important role in lowering these costs but put greater strain on the public sector or donor budgets.

Complementary interventions that either increase the effectiveness of CLTS or reduce the cost of implementation would make investments in these interventions more attractive. Sanitation marketing is one approach that could both increase uptake by focusing on demand generation and decrease latrine cost by supporting innovative technology (Devine & Kullmann 2011; Devine & Sijbesma 2011). However, these additional interventions, including other water, sanitation, or hygiene interventions, may be costly themselves and sometimes prove to be compliments or substitutes (Bennett 2012; Duflo *et al.* 2015). They would need to be analyzed to better understand their returns relative to other potential interventions (Abramovsky *et al.* 2019).

Moreover, researchers need to develop a stronger evidence base for understanding the effects of sanitation improvements and especially the nature of the positive externality from sanitation. The only health improvements included in the analysis presented here were those due to diarrhea reduction because data on other health effects were not available (Pearson & McPhedran 2008; Prüss-Ustün *et al.* 2014). Underscoring the importance of this issue, one-way sensitivity analysis revealed that the externality threshold assumption has a large effect on the predicted social net benefits of the CLTS plus subsidies intervention. The results in this analysis contrast with those in Radin *et al.* (2020), which found that the impact of the externality only had a modest impact on social net benefits from CLTS. The difference is due to differing assumptions about baseline conditions in that analysis and those in Ghana (including latrine coverage, diarrheal incidence, case fatality rates, and relevant costs). In principle, the herd protection could also be achieved by a CLTS-only intervention. That did not occur in this analysis because average baseline latrine coverage in rural Ghana is low and the latrine coverage increases documented in peer-reviewed impact evaluations of CLTS-only interventions were not sufficient to reach the threshold above which the externality was induced. Thus, only high-uptake communities targeted with CLTS plus subsidies achieve the level of community coverage necessary to generate the community externality. If this threshold was lower than assumed here or the additional decrease in diarrhea from herd protection was greater, the case for investment in CLTS-only interventions might improve.

This analysis demonstrates the importance of conducting economic analyses using context-specific data rather than simply assuming that global or regional analyses can help decision-makers prioritize local investments. To best support decision-making, future analyses should continue to incorporate real-world data that are specific to a region or local area.

ACKNOWLEDGEMENTS

The project was funded by the Copenhagen Consensus Center's Ghana Priorities Project. We would also like to thank the two anonymous reviewers for their helpful comments.

SUPPLEMENTARY MATERIAL

The Supplementary Material for this paper is available online at https://dx.doi.org/10.2166/washdev.2020.066.

REFERENCES

Abramovsky, L., Augsburg, B. & Oteiza, F. 2019 *Sustainable Total Sanitation-Nigeria: Final Research Report.* Report 156, Institute for Fiscal Studies, London, UK. Available from: https://www.ifs.org.uk/uploads/R156-Sustainable_Total_Sanitation%20%E2%80%93%20Nigeria.pdf (accessed 17 August 2020).

Addo, B. 2020 *Social Norms Advocacy Campaign on ODF Launched.* Ministry of Sanitation and Water Resources Republic of Ghana. Available from: http://mswr.gov.gh/social-norms-advocacy-campaign-on-odf-launched/ (accessed 12 August 2020).

Andres, L., Briceño, B., Chase, C. & Echenique, J. A. 2017 Sanitation and externalities: evidence from early childhood health in rural India. *Journal of Water, Sanitation and Hygiene for Development* **7** (2), 272–289. https://doi.org/10.2166/washdev.2017.143.

Augsburg, B., Caeyers, B. & Malde, B. 2019 *Can Micro-Credit Support Public Health Subsidy Programs?* Policy Research Working Paper No. 8846. World Bank Group Water Global Practice, Washington, DC. Available from: http://documents.worldbank.org/curated/en/424101557250234558/Can-Micro-Credit-Support-Public-Health-Subsidy-Programs (accessed 4 June 2020).

Banful, A. B. 2011 Old problems in the new solutions? Politically motivated allocation of program benefits and the 'new' fertilizer subsidies. *World Development* **39** (7), 1166–1176. doi:10.1016/j.worlddev.2010.11.004.

Barnard, S., Routray, P., Majorin, F., Peletz, R., Boisson, S., Sinha, A. & Clasen, T. 2013 Impact of Indian Total Sanitation Campaign on latrine coverage and use: a cross-sectional study in Orissa three years following programme implementation. *PLoS One* **8** (8), e71438. doi:10.1371/journal.pone.0071438.

Batmunkh, O., Chase, C., Galing, E. & Phuong La, M. 2019 *Final Evaluation Report on Integrating Sanitation Programming in the Pantawid Pamilya Program (Philippines)*. World Bank Group. Available from: http://pubdocs.worldbank.org/en/772751587563085022/Final-Evaluation-Report-on-Integrating-Sanitation-Programming-in-the-Philippines-Pantawid-Pamilya-Program.pdf (accessed 4 June 2020).

Bennett, D. 2012 Does clean water make you dirty? Water supply and sanitation in the Philippines. *Journal of Human Resources* **47** (1), 146–173. doi:10.3368/jhr.47.1.146.

Boisson, S., Sosai, P., Ray, S., Routray, P., Torondel, B., Schmidt, W., Bhanja, B. & Clasen, T. 2014 Promoting latrine construction and use in rural villages practicing open defecation: process evaluation in connection with a randomised controlled trial in Orissa, India. *BMC Research Notes* **7** (1). https://doi.org/10.1186/1756-0500-7-486.

Briceño, B., Coville, A., Gertler, P. & Martinez, S. 2017 Are there synergies from combining hygiene and sanitation promotion campaigns: evidence from a large-scale cluster-randomized trial in rural Tanzania. *PLoS One* **12** (11), e0186228. https://doi.org/10.1371/journal.pone.0186228.

Cameron, L., Shah, M. & Olivia, S. 2013 *Impact Evaluation of a Large-Scale Rural Sanitation Project in Indonesia*. The World Bank Water and Sanitation Program, Sustainable Development Network Policy Research Working Paper No. 6360.

Cameron, L., Olivia, S. & Shah, M. 2019 Scaling up sanitation: evidence from an RCT in Indonesia. *Journal of Development Economics* **138**, 1–16. https://doi.org/10.1016/j.jdeveco.2018.12.001.

Cavill, S., Chambers, R. & Vernon, N. 2015 *Sustainability and CLTS: Taking Stock. Frontiers of CLTS: Innovations and Insights*, Vol. 4. Institute of Development Studies, Unviersity of Sussex, Brighton, UK. Available from: https://www.communityledtotalsanitation.org/sites/communityledtotalsanitation.org/files/Frontiers4_Sustainability_0.pdf (accessed 10 September 2020)

Cha, S., Jung, S., Belew Bizuneh, D., Abera, T., Doh, Y.-A., Seong, J. & Ross, I. 2020 Benefits and costs of a community-led total sanitation intervention in rural Ethiopia – a trial-based ex post economic evaluation. *International Journal of Environmental Research and Public Health* **17** (14), 5068. https://doi.org/10.3390/ijerph17145068.

Clasen, T., Boisson, S., Routray, P., Torondel, B., Bell, M., Cumming, O., Ensink, J., Freeman, M., Jenkins, M. & Odagiri, M. 2014 Effectiveness of a rural sanitation programme on diarrhoea, soil-transmitted helminth infection, and child malnutrition in Odisha, India: a cluster-randomised trial. *The Lancet Global Health* **2** (11), e645–e653. https://doi.org/10.1016/S2214-109X(14)70307-9.

Coffey, D., Spears, D. & Vyas, S. 2017 Switching to sanitation: understanding latrine adoption in a representative panel of rural Indian households. *Social Science & Medicine* **188**, 41–50. https://doi.org/10.1016/j.socscimed.2017.07.001.

Crocker, J., Abodoo, E., Asamani, D., Domapielle, W., Gyapong, B. & Bartram, J. 2016a Impact evaluation of training natural leaders during a community-led total sanitation intervention: a cluster-randomized field trial in Ghana. *Environmental Science & Technology* **50** (16), 8867–8875. https://doi.org/10.1021/acs.est.6b01557.

Crocker, J., Geremew, A., Atalie, F., Yetie, M. & Bartram, J. 2016b Teachers and sanitation promotion: an assessment of community-led total sanitation in Ethiopia. *Environmental Science & Technology* **50** (12), 6517–6525. doi:10.1021/acs.est.6b01021.

Crocker, J., Saywell, D., Shields, K. F., Kolsky, P. & Bartram, J. 2017a The true costs of participatory sanitation: evidence from community-led total sanitation studies in Ghana and Ethiopia. *Science of the Total Environment* **601–602**, 1075–1083. https://doi.org/10.1016/j.scitotenv.2017.05.279.

Crocker, J., Saywell, D. & Bartram, J. 2017b Sustainability of community-led total sanitation outcomes: evidence from Ethiopia and Ghana. *International Journal of Hygiene and Environmental Health* **220** (3), 551–557. https://doi.org/10.1016/j.ijheh.2017.02.011.

Devereux, S., Masset, E., Sabates-Wheeler, R., Samson, M., Rivas, A. & Lintelo, D. 2017 The targeting effectiveness of social transfers. *Journal of Development Effectiveness* **9** (2), 162–211. https://doi.org/10.1080/19439342.2017.1305981.

Devine, J. & Kullmann, C. 2011 *Introductory Guide to Sanitation Marketing*. Report No. 72405, Water and Sanitation Program: Toolkit, World Bank, Water and Sanitation Program, Washington, DC. Available from: http://hdl.handle.net/10986/17352 (accessed 24 May 2020).

Devine, J. & Sijbesma, C. 2011 Sustainability of rural sanitation marketing in Vietnam: findings from a new case study. *Waterlines* **30** (1), 52–60. https://doi.org/10.3362/1756-3488.2011.005.

Duflo, E., Greenstone, M., Guiteras, R. & Clasen, T. 2015 *Toilets Can Work: Short and Medium Run Health Impacts of Addressing Complementarities and Externalities in Water and Sanitation*. Working Paper No. 21521, National Bureau of Economic Research, Cambridge. Available from: https://www.nber.org/papers/w21521.pdf (accessed 1 February 2020).

Elbers, C., Godfrey, S., Gunning, J. W., van der Velden, M. & Vigh, M. 2012 *Effectiveness of Large Scale Water and Sanitation Interventions: The One Million Initiative in Mozambique*. Tinbergen Institute Discussion Papers, 12–069/2, Tinbergen Institute, Amsterdam.

Fuller, J. A., Villamor, E., Cevallos, W., Trostle, J. & Eisenberg, J. N. 2016 I get height with a little help from my friends: herd protection from sanitation on child growth in rural Ecuador. *International Journal of Epidemiology* **45** (2), 460–469. https://doi.org/10.1093/ije/dyv368.

Garn, J. V., Sclar, G. D., Freeman, M. C., Penakalapati, G., Alexander, K. T., Brooks, P., Rehfuess, E. A., Boisson, S., Medlicott, K. O. & Clasen, T. F. 2017 The impact of sanitation interventions on latrine coverage and latrine use: a systematic review and meta-analysis. *International Journal of Hygiene and Environmental Health* **220** (2), 329–340. https://doi.org/10.1016/j.ijheh.2016.10.001.

Ghana Statistical Service, Ghana Health Service, & ICF 2018 *Ghana Maternal Health Survey 2017: Key Findings*. Ghana Statistical Service, Ghana Health Service, and ICF, Rockville, MD. Available from: https://www.dhsprogram.com/pubs/pdf/SR251/SR251.pdf (accessed 1 February 2020).

Global Burden of Disease Collaborative Network 2017 *Global Burden of Disease Study 2016 (GBD 2016) Results*. Institute for Health Metrics and Evaluation (IHME), Seattle, WA. Available from: http://ghdx healthdata.org/gbd-results-tool (accessed 1 February 2020).

Guiteras, R., Levinsohn, J. & Mobarak, A. M. 2015 Encouraging sanitation investment in the developing world: a cluster-randomized trial. *Science* **348** (6237), 903–906. https://doi.org/10.1126/science.aaa0491.

Hammer, J. & Spears, D. 2016 Village sanitation and children's human capital: evidence from a randomized experiment by the Maharashtra government. *Journal of Health Economics* **48**, 135–148. https://dx.doi.org/10.1016/j.jhealeco.2016.03.003.

Harris, M., Alzua, M. L., Osbert, N. & Pickering, A. 2017 Community-level sanitation coverage more strongly associated with child growth and household drinking water quality than access to a private toilet in rural Mali. *Environmental Science & Technology* **51** (12), 7219–7227. https://doi.org/10.1021/acs.est.7b00178.

Heijnen, M., Cumming, O., Peletz, R., Chan, G. K.-S., Brown, J., Baker, K. & Clasen, T. 2014 Shared sanitation versus individual household latrines: a systematic review of health outcomes. *PLoS One* **9** (4), e93300. https://doi.org/10.1371/journal.pone.0093300.

Humphrey, J. H., Mbuya, M. N., Ntozini, R., Moulton, L. H., Stoltzfus, R. J., Tavengwa, N. V., Mutasa, K., Majo, F., Mutasa, B., Mangwadu, G., Chasokela, C. M., Chigumira, A., Chasekwa, B., Smith, L. E., Tielsch, J. M., Jones, A. D., Manges, A. R., Maluccio, J. A. & Prendergast, A. J. 2019 Independent and combined effects of improved water, sanitation, and hygiene, and improved complementary feeding, on child stunting and anaemia in rural Zimbabwe: a cluster-randomised trial. *The Lancet Global Health* **7** (1), e132–e147. https://doi.org/10.1093/cid/ciy844.

Israel, M. & Shapiro, J. 2019 *Study of Targeted Subsidies Within ODF Communities in Ghana*. USAID Water, Sanitation, and Hygiene Partnerships for Learning and Sustainability Project (USAID/WASHPaLS). Available from: https://www.globalwaters.org/resources/assets/washpals/study-targeted-subsidies-within-odf-communities-ghana (accessed 28 May 2020).

Jenkins, M. & Sugden, S. 2006 *Rethinking Sanitation – Lessons and Innovation for Sustainability and Success in the New Millennium*. United Nations Development Program. Available from: http://hdr.undp.org/en/content/rethinking-sanitation-lessons-and-innovation-sustainability-and-success-new-millennium (accessed 28 May 2020).

Jung, Y. T., Hum, R. J., Lou, W. & Cheng, Y. L. 2017 Effects of neighbourhood and household sanitation conditions on diarrhea morbidity: systematic review and meta-analysis. *PLoS One* **12** (3), e0173808. https://doi.org/10.1371/journal.pone.0173808.

Kar, K. & Chambers, R. 2008 *Handbook on Community-Led Total Sanitation*. Plan UK, London, UK. Available from: https://www.communityledtotalsanitation.org/sites/communityledtotalsanitation.org/files/cltshandbook.pdf (accessed 28 May 2020).

Luby, S. P., Rahman, M., Arnold, B. F., Unicomb, L., Ashraf, S., Winch, P. J., Stewart, C. P., Begum, F., Hussain, F., Benjamin-Chung, J., Leontsini, E., Naser, A. M., Parvez, S. M., Hubbard, A. E., Lin, A., Nizame, F. A., Jannat, K., Ecrumen, A., Ram, P. K., Das, K. K., Abedin, J., Clasen, T. F., Dewey, K. G., Fernald, L. C., Null, C., Ahmed, T. & Colford Jr., J. M. 2018 Effects of water quality, sanitation, handwashing, and nutritional interventions on diarrhoea and child growth in rural Bangladesh: a cluster randomised controlled trial. *The Lancet Global Health* **6** (3), e302–e315. http://dx.doi.org/10.1016/S2214-109X(17)30490-4.

Ministry of Sanitation and Water Resources 2016 *Guidelines for Targeting the Poor and Vulnerable for Basic Sanitation Services in Ghana*. Government of Ghana, Accra.

National Development Planning Commission 2017 *Long-term National Development Plan of Ghana (2018–2057) Highlights*. National Development Planning Commission, Accra, Ghana.

Nonvignon, J., Atherly, D., Pecenka, C., Aikins, M., Gazley, L., Groman, D., Narh, C. T. & Armah, G. 2018 Cost-effectiveness of rotavirus vaccination in Ghana: examining impacts from 2012 to 2031. *Vaccine* **36** (47), 7215–7221. https://doi.org/10.1016/j.vaccine.2017.11.080.

Null, C., Stewart, C. P., Pickering, A. J., Dentz, H. N., Arnold, B. F., Arnold, C. D., Benjamin-Chung, J., Clasen, T., Dewey, K. G., Fernald, L. C., Hubbard, A. E., Kariger, P., Lin, A., Luby, S. P., Mertens, A., Njenga, S. M., Nyambane, G., Ram, P. K. & Colford Jr., J. M. 2018 Effects of water quality, sanitation, handwashing, and nutritional interventions on diarrhoea and child growth in rural Kenya: a cluster-randomised controlled trial. *The Lancet Global Health* **6** (3), e316–e329. https://doi.org/10.1016/S2214-109X(18)30005-6.

ODF Ghana Campaign 2017 Available from: http://odfghanacampaign.org/ (accessed 4 November 2019).

Orgill-Meyer, J., Pattanayak, S. K., Chindarkar, N., Dickinson, K. L., Panda, U., Rai, S., Sahoo, B., Singha, A. & Jeuland, M. 2019 Long-term impact of a community-led sanitation campaign in India, 2005–2016. *Bulletin of the World Health Organization* **97** (8), 523–533A. https://doi.org/10.2471/BLT.18.221572.

Palermo, T. M., Valli, E., Ángeles-Tagliaferro, G., de Milliano, M., Adamba, C., Spadafora, T. R. & Barrington, C. 2019 Impact evaluation of a social protection programme paired with fee waivers on enrolment in Ghana's national health insurance scheme. *BMJ Open* **9** (11). http://dx.doi.org/10.1136/bmjopen-2018-028726.

Patil, S. R., Arnold, B. F., Salvatore, A. L., Briceno, B., Ganguly, S., Colford, J. M. & Gertler, P. J. 2014 The effect of India's total sanitation campaign on defecation behaviors and child health in rural Madhya Pradesh: a cluster randomized controlled trial. *PLoS Medicine* **11** (8). https://doi.org/10.1371/journal.pmed.1001709.

Pattanayak, S. K., Yang, J., Dickinson, K. L., Poulos, C., Patil, S. R., Mallick, R. K., Blitstein, J. L. & Praharaj, P. 2009 Shame or subsidy revisited: social mobilization for sanitation in Orissa, India. *Bulletin of the World Health Organization* **87** (8), 580–587. https://doi.org/10.2471/BLT.08.057422.

Pearson, J. & McPhedran, K. 2008 A literature review of the non-health impacts of sanitation. *Waterlines* **27** (1), 1756–3488. https://doi.org/10.3362/1756-3488.2008.005.

Peletz, R., Cock-Esteb, A., Ysenburg, D., Haji, S., Khush, R. & Dupas, P. 2017 Supply and demand for improved sanitation: results from randomized pricing experiments in rural Tanzania. *Environmental Science & Technology* **51** (12), 7138–7147. doi:10.1021/acs.est.6b03846.

Peletz, R., Kisiangani, J., Ronoh, P., Cock-Esteb, A., Chase, C., Khush, R. & Luoto, J. 2019 Assessing the demand for plastic latrine slabs in rural Kenya. *The American Journal of Tropical Medicine and Hygiene* **101** (3), 555–565. doi:10.4269/ajtmh.18-0888.

Pickering, A. J., Djebbari, H., Lopez, C., Coulibaly, M. & Alzua, M. L. 2015 Effect of a community-led sanitation intervention on child diarrhoea and child growth in rural Mali: a cluster-randomised controlled trial. *The Lancet Global Health* **3** (11), e701–e711. https://doi.org/10.1016/S2214-109X(15)00144-8.

Prüss-Ustün, A., Bartram, J., Clasen, T., Colford Jr, J. M., Cumming, O., Curtis, V., Bonjour, S., Dangour, A. D., De France, J., Fewtrell, L. & Freeman, M. C. 2014 Burden of disease from inadequate water, sanitation and hygiene in low-and middle-income settings: a retrospective analysis of data from 145 countries. *Tropical Medicine & International Health* **19** (8), 894–905. https://doi.org/10.1111/tmi.12329.

Radin, M., Jeuland, M., Wang, H. & Whittington, D. 2020 Benefit-cost analysis of community-led total sanitation: incorporating results from recent evaluations. *Journal of Benefit-Cost Analysis*. https://doi.org/10.1017/bca.2020.6.

Riahi, K., Van Vuuren, D. P., Kriegler, E., Edmonds, J., O'Neill, B. C., Fujimori, S., Bauer, N., Calvin, K., Dellink, R., Fricko, O., Lutz, W., Popp, A., Cauresma, J. C., Samir, K. C., Leimach, M., Jiang, L., Kram, T., Rao, S., Emmerling, J., Ebi, K., Hasegawa, T., Havlík, P., Humpenöder, F., Da Silva, L. A., Smith, S., Stehfest, E., Bosetti, V., Eom, J., Gernaat, D., Masui, T., Rogelj, J., Strefler, J., Drouet, L., Krey, V., Luderer, G., Harmsen, M., Takahashi, K., Baumstark, L., Doelman, J. C., Kainuma, M., Klimont, Z., Marangoni, G., Lotze-Campen, H., Obersteiner, M., Tabeau, A. & Tavoni, M. 2017 The shared socioeconomic pathways and their energy, land use, and greenhouse gas emissions implications: an overview. *Global Environmental Change* **42**, 153–168. https://doi.org/10.1016/j.gloenvcha.2016.05.009.

Robinson, L. A. & Hammitt, J. K. 2018 *Valuing Nonfatal Health Risk Reductions in Global Benefit-Cost Analysis*. Working Paper No. 2, Guidelines for Benefit-Cost Analysis Project, Center for Health Decision Science, Harvard TH Chan School of Public Health, Boston, MA. Available from: https://cdn2.sph.harvard.edu/wp-content/uploads/sites/94/2017/01/Robinson-Hammitt-Nonfatal-Risks.2018.03.121.pdf (accessed 28 May 2020).

Robinson, L. A., Hammitt, J. K., Cecchini, M., Chalkidou, K., Claxton, K., Cropper, M., Eozenou, P. H., de Ferranti, D., Deolalikar, A. B., Guanais, F., Jamison, D. T., Kwon, S., Lauer, J. A., O'Keeffe, L., Walker, D., Whittington, D., Wilkinson, T., Wilson, D. & Wong, B. 2019a *Reference Case Guidelines for Benefit-Cost Analysis in Global Health and Development*. Center for Health Decision Science, Harvard TH Chan School of Public Health, Boston, MA. Available from: https://cdn1.sph.harvard.edu/wp-content/uploads/sites/2447/2019/05/BCA-Guidelines-May-2019.pdf (accessed 28 May 2020).

Robinson, L. A., Hammitt, J. K. & O'Keeffe, L. 2019b Valuing mortality risk reductions in global benefit-cost analysis. *Journal of Benefit-Cost Analysis* **10** (S1), 15–50. https://doi.org/10.1017/bca.2018.26.

Samir, K. C. & Lutz, W. 2017 The human core of the shared socioeconomic pathways: population scenarios by age, sex and level of education for all countries to 2100. *Global Environmental Change* **42**, 181–192. https://doi.org/10.1016/j.gloenvcha.2014.06.004.

SNV 2010 *SNV Vietnam Sanitation Demand Creation – A Strengths-Based Review*. SNV, Vietnam. Available from: https://www.Ircwash.org/Sites/Default/Files/snv_Vietnam_Clts_Strengths_Based_Review_Report_dec2010.pdf (accessed 4 June 2020).

The World Bank 2011 *Ghana Looks to Retool Its Economy as It Reaches Middle-Income Status*. Available from: https://www.worldbank.org/en/news/feature/2011/07/18/ghana-looks-to-retool-its-economy-as-it-reaches-middle-income-status (accessed 14 September 2020).

UNICEF 2017 *Country Office Annual Report 2017 – Ghana*. Accra, Ghana. Available from: https://www.unicef.org/about/annualreport/files/Ghana_2017_COAR.PDF (accessed 28 May 2020).

Venkataramanan, V. 2014 *Testing CLTS Approaches for Scalability. CLTS Learning Series: Cambodia PDR Country Report*. Plan Institute USA Inc., The Water Institute, Gillings School of Global Public Health, The University of North Carolina, Chapel Hill, NC. Available from: https://waterinstitute.unc.edu/files/2015/09/learning-series-cambodia-country-report-2015-08_updated.pdf (accessed 4 June 2020).

Whittington, D. & Cook, J. 2019 Valuing changes in time use in low-and middle-income countries. *Journal of Benefit-Cost Analysis* **10** (S1), 51–72. https://doi.org/10.1017/bca.2018.21.

Whittington, D., Hanemann, M. W., Sadoff, C. & Jeuland, M. 2009 The challenge of improving water and sanitation services in less developed countries. *Foundations and Trends in Microeconomics* **4** (6–7), 469–609. https://doi.org/10.1561/0700000030.

Whittington, D., Radin, M. & Jeuland, M. 2020 Evidence-based policy analysis? The strange case of the randomized controlled trials of community-led total sanitation. *Oxford Review of Economic Policy* **36** (1), 191–221. https://doi.org/10.1093/oxrep/grz029.

Wolf, J., Hunter, P. R., Freeman, M. C., Cumming, O., Clasen, T. F., Bartram, J., Higgins, J. P. T., Johnston, R., Medlicott, K. & Boisson, S. 2018 Impact of drinking water, sanitation and handwashing with soap on childhood diarrhoeal disease: updated meta-analysis and meta-regression. *Tropical Medicine & International Health* **23** (5), 508–525. https://doi.org/10.1111/tmi.13051.

World Health Organization & United Nation's Children Fund Joint Monitoring Program 2019 *Progress on Household Drinking Water, Sanitation and Hygiene 2000–2017. Special Focus on Inequalities*. United Nations Children's Fund & World Health Organization, New York. Available from: https://www.unicef.org/reports/progress-on-drinking-water-sanitation-and-hygiene-20191 (accessed 28 May 2020).

World Health Organization & United Nation's Children Fund Joint Monitoring Program 2020 *JMP Household Data Tables*. Available from: https://washdata.org/data/household#!/table?geo0=country&geo1=GHA (accessed 28 May 2020).

Wright, J., Dzodzomenyo, M., Fink, G., Wardrop, N. A., Aryeetey, G. C., Adanu, R. M. & Hill, A. G. 2016 Subsidized sachet water to reduce diarrheal disease in young children: a feasibility study in Accra, Ghana. *The American Journal of Tropical Medicine and Hygiene* **95** (1), 239–246. https://doi.org/10.4269/ajtmh.15-0854.

Xignite 2018 *USD to GHS Rates on 1/1/2018*. Available from: https://www.exchange-rates.org/Rate/USD/GHS/1-1-2018 (accessed 28 May 2020).

Zuin, V., Delaire, C., Peletz, R., Cock-Esteb, A., Khush, R. & Albert, J. 2019 Policy diffusion in the rural sanitation sector: lessons from community-led total sanitation (CLTS). *World Development* **124** (104643), 1–14. https://doi.org/10.1016/j.worlddev.2019.104643.

First received 5 March 2020; accepted in revised form 12 October 2020. Available online 27 October 2020

Research Paper

Expanding safe fecal sludge management in Kisumu, Kenya: an experimental comparison of latrine pit-emptying services

Rachel Peletz, Andy Feng, Clara MacLeod, Dianne Vernon, Tim Wang, Joan Kones, Caroline Delaire, Salim Haji and Ranjiv Khush

ABSTRACT

Most residents of Kisumu, Kenya, use latrines constructed over basic pits or attached to more durable concrete vaults and septic tanks. Only one-third of fecal sludge generated in the city, however, is safely collected and treated. Programs for improving fecal sludge management among poor households include the development of formal manual emptying organizations that are recognized by local authorities, employ safety procedures, and transport fecal sludge to a treatment site. In this study, we compared the financial structures of these organizations with those of vacuum trucks that primarily serve wealthier households. We also employed an incentives-based strategy to promote the expansion of safe pit-emptying services in a low-income area and compared the performance of three managing groups to coordinate these services: (1) The Association of Wastewater Managers (The Association); (2) a formal manual emptying organization; and (3) a community-based water supplier interested in coordinating emptying services. Vacuum trucks were more cost-effective than the formal manual emptying organization, and The Association was most efficient in servicing poor households. The Association also demonstrated the ability to service low-income areas comprehensively by delegating a fraction of jobs (11%) to formal manual emptiers in locations not serviceable by VTOs, and overall showed the highest potential to achieve pro-poor service delivery at scale.

Key words | fecal sludge management, Kenya, latrines, low-income, sanitation, urban

Rachel Peletz
Caroline Delaire
Salim Haji
Ranjiv Khush (corresponding author)
The Aquaya Institute,
P.O. Box 1603, San Anselmo, CA 94979,
USA
E-mail: *ranjiv@aquaya.org*

Andy Feng
Oliver Wyman,
New York, NY 10036,
USA

Clara MacLeod
Joan Kones
The Aquaya Institute,
P.O. Box 21862-00505, Nairobi,
Kenya

Dianne Vernon
Oliver Wyman,
San Francisco, CA 94111,
USA

Tim Wang
Oliver Wyman,
Toronto, ON M5 J 0A1,
Canada

HIGHLIGHTS

- Sanitation conditions are of increasing concern for rapidly growing cities of the developing world.
- In Kisumu, Kenya, most residents use latrines constructed over basic pits or attached to more durable concrete vaults and septic tanks. However, only one-third of the fecal sludge generated in the city is safely collected and treated.
- Efforts to improve fecal sludge management among low-income households in Kisumu include the development of formal manual emptying organizations that are recognized by local

doi: 10.2166/washdev.2020.060

authorities, employ safety procedures and equipment, and transport fecal sludge to the municipal treatment site.

- This study compares the financial structures of these formal manual emptying organizations with those of vacuum trucks that primarily serve wealthier households to determine the feasibility of expanding safe fecal sludge services to low-income areas in Kisumu.
- We employed an incentives-based strategy to compare pit-emptying service provision in a low-income area by three different groups, which include vacuum trucks and formal manual emptiers. We determined that vacuum trucks were more cost-effective than the formal manual emptying organization, and the Association of VTOs was most efficient in providing emptying services to poor households.

INTRODUCTION

Located in Western Kenya and bordering Lake Victoria, Kisumu is the country's third-largest city with an estimated 419,000 inhabitants, of which approximately 60% live in low-income areas (County Government of Kisumu 2018a; Furlong & Jooust 2016). As is the case for many cities in the developing world, sewerage networks are limited in Kisumu, serving only 20% of households. Approximately 70% of households rely on latrines constructed over simple pits and about 5% of households, predominantly in middle- and high-income areas, use latrines linked to underground concrete vaults or septic tanks (Furlong & Jooust 2016). It is estimated that the remaining 5% practice open defecation. These underground containment structures are typically either emptied by vacuum truck operators (VTOs), who mostly serve wealthier households, or informal manual emptiers who remove fecal sludge by hand or buckets and bury it onsite or dispose of it in nearby waterways. Estimates indicate that only one-third of Kisumu's collected fecal sludge is treated at the city's sewage treatment sites, which are managed by Kisumu Water and Sanitation Company (KIWASCO); the remainder is directly discharged into the environment or buried (Furlong & Jooust 2016). Increasing the use of safe and regulated emptying services is therefore critical for improving sanitation in Kisumu.

Recent efforts by the municipality and development partners to promote better fecal sludge management in Kisumu, particularly in low-income areas, have largely focused on the formation of authorized manual emptying services that adhere to safety practices and utilize authorized treatment sites. In 2016, the Kisumu City Partnership for Improved Sanitation in Informal Settlements (KisumuSan), led by the international non-profit organization, Practical Action, trained two community-based organizations, BlueStars and Vukasasa, in safe manual emptying (Practical Action 2016). Prior to joining the KisumuSan program, BlueStars members were informal pit emptiers and Vukasasa members were dedicated to cleaning public and school toilet facilities, cleaning school compounds, and managing an ablution block. In 2018, Water and Sanitation for the Urban Poor (WSUP), an international not-for-profit, implementing organization began collaborating with the Kisumu County Public Health Office to develop Standard Operating Procedures (SOPs) for fecal sludge management (Owako & Renouf 2018). These SOPs provide guidance on emptying practices and equipment that were designed to protect the health of manual emptiers. The SOPs also specify safe methods for transporting and disposing of fecal sludge. In addition, WSUP, in partnership with the Kisumu County Public Health Office and KIWASCO, helped a solid waste management company (Gasia Poa) expand into fecal sludge management (Owako & Renouf 2018; WSUP 2018). WSUP provided Gasia Poa with training in the SOPs and guidance in marketing, branding, and financial management. The three emptying organizations (BlueStars, Vukasasa, and Gasia Poa) now operate in various low-income areas of Kisumu. We refer to these organizations as 'formal' manual emptiers: they are authorized by the Kisumu County Government and the National

Environment Management Authority (NEMA), and they are allowed to dump waste at the KIWASCO fecal sludge treatment site without paying a dumping fee. They typically transfer fecal sludge into barrels (using the hand-operated Gulper mechanical pump or steel buckets with long handles) and then transport the barrels in a pick-up truck to the treatment site.

The promotion of formalized manual emptying services in Kisumu and other cities of the developing world is generally based on assumptions that vacuum trucks are not appropriate for low-income areas: it is generally believed that their services are too expensive for poor households, they cannot access households in dense urban settlements, and they are not equipped to remove the solid waste typically found in pit latrines (Chowdhury & Kone 2012; Balasubramanya *et al.* 2017; Mansour Oyaya & Owor 2017). To test these assumptions, we undertook two research activities: (1) we determined the costs of pit-emptying by vacuum trucks and formal manual emptying organizations; and (2) we established an incentives-based latrine pit-emptying program to compare the performance of three managing groups in coordinating the delivery of emptying services in a low-income area of Kisumu.

METHODS

Landscape assessment

We began our comparisons of latrine pit-emptying options in Kisumu by establishing the landscape of existing emptying services: VTOs, formal manual emptying organizations (BlueStars, Vukasasa, Gasia Poa), and informal manual emptiers (Figure 1). To verify business models, establish customer bases, and identify barriers for expansion within low-income areas, we conducted 18 interviews and observations of pit-emptying service providers. Additionally, we reviewed nine policy documents and 25 reports on past sanitation programs and interviewed 17 government and non-governmental stakeholders to understand the regulatory environment (Figure 1). To understand efforts to improve sanitation in other Kenyan cities, we interviewed sanitation service providers in Nakuru (Nakuru Water and Sanitation Services Company), Malindi (Malindi Water and Sewerage Company), and Nairobi (Sanergy, a fecal sludge management organization with for-profit and non-profit operations). Finally, to establish a forum for regularly engaging with local stakeholders, we formed a working committee in Kisumu that included the City and County Public Health Departments, KIWASCO, Great Lakes University Kisumu, and WSUP. The research team met five times with the working committee over the course of the study.

Cost of safe pit-emptying

We used the information collected during interviews with service providers to estimate the costs that they incur per emptying job. We considered both capital expenditures (CAPEX) and operational expenditures (OPEX). For the group of formal manual emptiers, CAPEX included one Gulper (estimated lifetime: 10 years), six 50-L barrels (estimated lifetime: 5 years), and five sets of personal protective equipment (gumboots, helmet, overalls, mask, gloves) with an estimated lifetime of 1–2 years. For a VTO, CAPEX included an 8,000-L exhauster truck (estimated lifetime: 10 years), three 100-feet hoses (estimated lifetime: 1 year), one set of personal protective equipment (gumboots, overalls, gloves) with an estimated lifetime of 1–6 months, annual licenses to the municipality and NEMA, and a one-time registration with the Association of Wastewater Managers. OPEX included labor, truck rental, office costs, and disinfectant consumables for formal manual emptiers; labor, truck maintenance, fuel, a monthly permit to access the treatment facility, and a quarterly membership fee to the Association for VTOs. To estimate costs per emptying job, we first calculated annual costs and then normalized by the approximate annual number of jobs: 2,160 for a VTO (or 6 jobs per day) and 60 for the group of manual emptiers (5 jobs per month), based on currently reported levels of activity.

Managing groups for pit-emptying

Our sanitation landscape assessment and working committee stakeholder consultations identified the need for exploring different management models for pit-emptying services. For our comparative study, we applied the following criteria to select groups that were either already

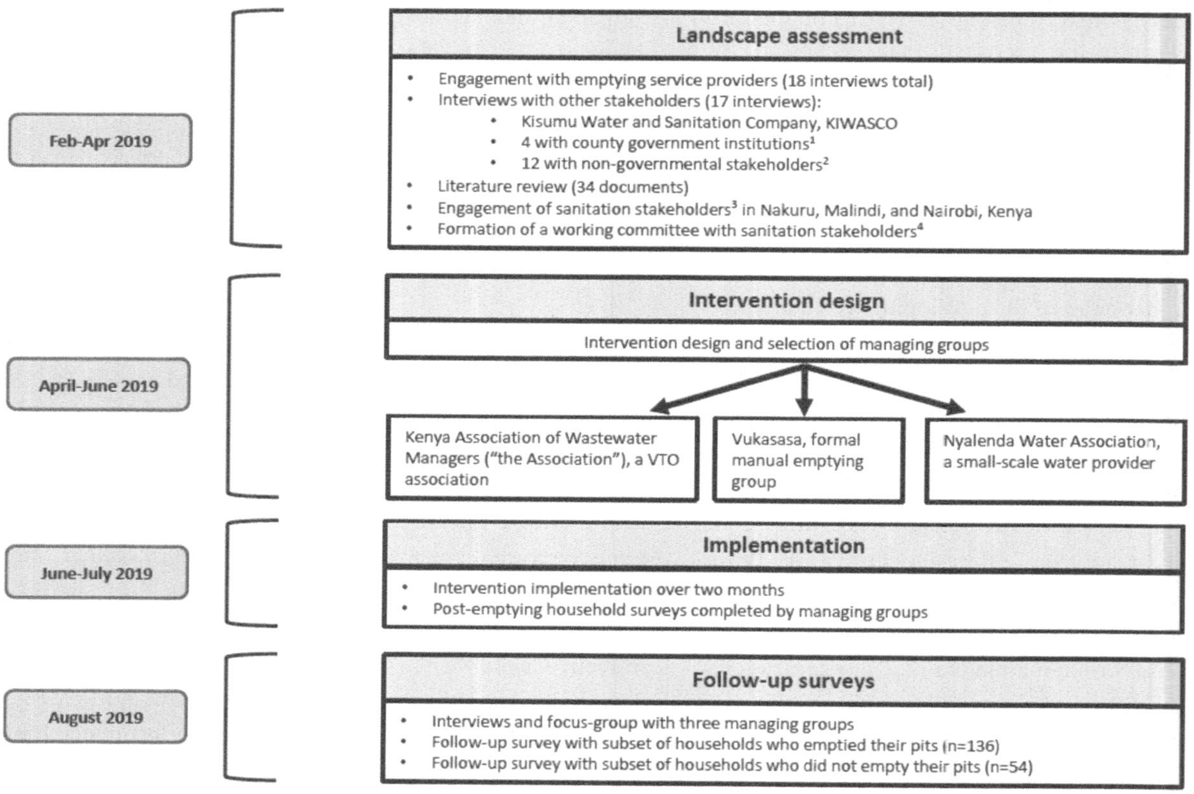

Figure 1 | Study flow diagram. [1]County government institutions include: Kisumu County Public Health Office, Kisumu City Public Health Office, National Environmental Management Authority, and Lake Victoria South Water Service Board. [2]Non-governmental stakeholders include: the Kenya Integrated Water, Sanitation, and Hygiene (KIWASH) program funded by USAID, Kenya Women Finance Trust, ECLOF Kenya (microfinance institution), Kenya Urban Apostolate Program, Umande Trust, Practical Action, Water and Sanitation for the Urban Poor, and Great Lakes University Kisumu. [3]Sanitation stakeholders included Nakuru Water and Sanitation Services Company in Nakuru, Malindi Water and Sewerage Company in Malindi, and the social enterprise Sanergy in Nairobi. [4]The working committee consisted of Kisumu City and County Public Health Departments, Kisumu Water and Sanitation Company (KIWASCO), Great Lakes University Kisumu, Water and Sanitation for the Urban Poor (WSUP).

providing safe pit-emptying services or interested in expanding into pit-emptying: (i) currently operating in the low-income area of Nyalenda within Kisumu, (ii) established business expertise (e.g., experience providing services and managing money), (iii) willingness to participate in a research study, and (iv) concurrence from the working committee. This selection process identified three groups with different cost and management structures: (1) Kenya Association of Wastewater Managers (The Wastewater Association); (2) Vukasasa, the primary formal manual emptying organization operating in Nyalenda; and (3) the Nyalenda Water Association (Nyalenda Water), one of KIWASCO's small-scale water providers responsible for distributing piped water in Kisumu's low-income areas through a delegated management model (Castro 2009; Nzengya 2015; Schwartz & Sanga 2015) (Table 1). Throughout our comparative study, these selected groups managed their delivery of safe emptying services: they were responsible for marketing services, identifying customers, setting prices, and coordinating pit-emptying jobs. All three managing groups could delegate jobs to either VTOs or formal manual emptiers. In the case of the Wastewater Association, delegating a job to a VTO meant to assign it to one of its own members. Similarly, for Vukasasa, delegating a job to formal manual emptiers meant to assign it to its own members.

To standardize operating conditions for the three managing groups during our study, we selected three separate geographic zones in the Nyalenda area (out of 12 zones that we evaluated) according to the following specifications: (i) inclusion of 500–700 households; (ii) equidistant to the KIWASCO treatment site; (iii) demarcated by clear

Table 1 | Description of the three managing groups selected for the comparative study of pit-emptying

Managing group	Description and structure
The Kenya Association of Wastewater Managers *(referred to as 'The Wastewater Association')*	• An association of VTOs • Serves as an intermediary to link customers to VTOs • Subcontracts jobs to formal manual emptying groups as needed • VTOs operate in all of Kisumu
Vukasasa	• A community-based organization that provides safe manual emptying services • Able to subcontract jobs to VTOs, though rarely applied in practice • Only operates in the low-income area 'Nyalenda'
Nyalenda Water Association *(referred to as 'Nyalenda Water')*	• A small-scale water supplier that provides water in low-income areas • Overseen by Kisumu Water and Sewerage Company (KIWASCO) through a delegated management model • Only operates in the low-income area 'Nyalenda' • No previous experience in sanitation and, therefore, manages pit-emptying by subcontracting jobs to VTOs or formal manual emptiers

boundaries such as roads or rivers; (iv) near the offices of the managing group (did not apply to the Wastewater Association); and (v) approved by the managing groups.

Cash incentives

To encourage participation in our comparative study of pit-emptying services, we provided cash incentives to the managing groups (Figure 2). In addition to the payments that they negotiated with customers, we provided managing groups with a 3,000 KES (30 USD) cash incentive for each emptying job that they completed within their assigned zone during our study period We selected this amount to address the estimated gap between market prices for VTO and formal manual pit-emptying services in Kisumu and the average willingness-to-pay (WTP) for pit-emptying among poor households, who largely rely on informal manual emptiers (who charge 10–30 USD per job according to our landscape assessment). We measured WTP in a parallel study (Peletz *et al.*, forthcoming). Recognizing that the costs of formal manual emptying were higher than VTOs primarily due to transport, we provided an additional incentive payment of 2,000 KES (20 USD) per trip to the treatment site for pit-emptying jobs completed by formal manual emptiers to support the costs of transporting fecal sludge from the study zones to the KIWASCO treatment site. These incentives were necessary to promote the expansion of formal pit-emptying services into previously

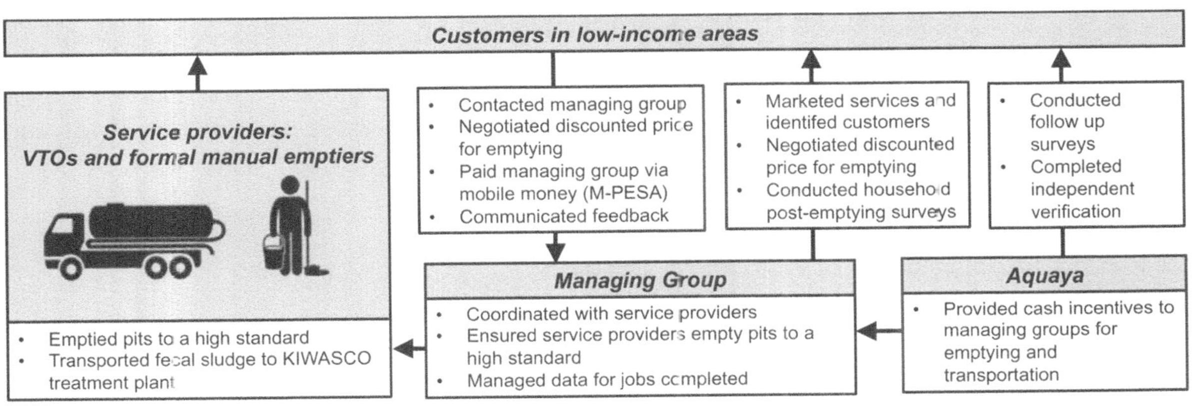

Figure 2 | Comparative study design. Each group's responsibilities are represented by the direction of the arrows. The research team (Aquaya) provided cash incentives to managing groups for emptying and transport.

non-serviced areas. We implemented them for a period of 2 months, as defined by our available budget. This period was sufficient to measure differences in performance between the three managing groups.

Other than defining the zones and providing cash incentives as described above, we did not apply other constraints or guidelines on the three groups. We did not provide capacity building to any of the managing groups. A fundamental element of the research design was giving the groups substantial flexibility to determine key elements of their business model, including choice of provider (VTO or manual), price charged to customer (reflecting the specifics of the job including condition and accessibility of pit), and the price paid to the service provider.

Data collection

We collected data from June to July 2019 to assess the following outcome metrics for each managing group: (i) *safe emptying performance* (i.e., number of households served and volume of fecal waste safely removed), (ii) *financial performance* of managing groups (e.g., prices charged to customers, costs, and net revenues), and (iii) *customer satisfaction*. To capture data electronically, we provided one smart phone (Samsung Galaxy J4, South Korea) carrying the CommCare survey and data management application (DiMagi Inc., Cambridge, MA, USA) to each managing group. After each emptying job, managing groups completed a post-emptying survey to document the toilet type, number of users, volume of sludge removed, number of trips required to transport sludge to the treatment site, customer satisfaction, payment received from the customer, and any payment to a subcontracted emptier. To qualify for our incentive payments, the managing group was required to submit a photo of each emptied pit and receipts that documented their financial transactions with customers and subcontractors, if applicable. We also hired an independent field coordinator to verify 10% of emptying jobs for each managing group, with a minimum of one verification per week per group.

To audit emptying records and customer feedback reports, we attempted follow-up surveys with all of the customers serviced by Vukasasa (60 households) and Nyalenda Water (44 households) and a random selection of 60% (81/136) of the households serviced by The Wastewater Association. To understand why households did not use the safe emptying services, we conducted follow-up surveys with 54 randomly selected households that did not have their pits emptied by managing groups. We entered household survey responses into the CommCare survey and data management application (DiMagi Inc., Cambridge, MA, USA) on mobile phones (Samsung Galaxy J4, South Korea). We also conducted individual interviews with each managing group and a combined focus group discussion with all three groups to discuss successes, challenges, and recommendations for expanding their businesses in low-income areas.

Data analysis

We analyzed data using the spreadsheet software program Microsoft Excel (Microsoft Corp. Redmond, WA, USA) to examine key outcome metrics previously described. We randomly selected households for follow-up surveys using Excel (for those that received services) and ArcGIS (for those that had not received services, identified using randomly generated GPS points). The exchange rate that we used during our analysis was 1.00 USD = 100 KES (6 June 2019, oanda.com).

Ethical approval

The Western Institutional Review Board (WIRB) (Puyallup, WA, USA) determined that this study was exempt from full ethical review under 45 CFR §46.101(b)(2) of the Federal Common Rule in the USA. In Kenya, we obtained ethical approval for our research from Amref Health Africa (AMREF) (ESRC P493/2018) and a research permit from the National Commission for Science, Technology, and Innovation (NACOSTI) (NACOSTI/P/19/39980/28701).

RESULTS

Costs of safe emptying by different service providers

We analyzed the capital and operating expenses of VTOs and formal manual emptying organizations in Kisumu prior to beginning our comparative study with the three managing groups (Table 2). Our analyses showed that per emptying job, formal manual emptying was about four times more expensive (6,169 KES or 62 USD) than emptying by VTOs (1,630 KES or 16 USD). Key drivers of this difference were the higher OPEX incurred by formal manual emptiers for labor (25.5 USD per job compared to 11.2 USD per job for VTOs) and for transporting fecal sludge to the treatment site (30 USD per job compared to 3.7 USD per job for VTOs). Economies of scale further improved the cost-effectiveness of VTOs by lowering per-job CAPEX (formal emptying groups performed an average of five pits per month, compared to six jobs per day for VTOs).

Data accountability and misrepresentation

During our comparative study, we found that two of the managing groups misrepresented data and mismanaged funds, despite our efforts to hire an independent field coordinator to verify emptying jobs during the comparative study. Based on our follow-up audit, we estimated that 32% (19/60) of Vukasasa's pit-emptying records and 59% (26/44) of Nyalenda Water's pit-emptying records were invalid: in these cases, households either could not be re-located (according to the recorded GPS coordinates and phone numbers) or did not have their pits emptied. The two managing groups did not provide valid explanations for why, in some cases, we could not re-locate households nor why, in other cases, pits that were not emptied were reported as emptied. In contrast, we determined that only 1% of the Wastewater Association's records (1/81) were invalid. We disregarded the pit-emptying jobs that we could not confirm and reported the total number of jobs accordingly (Table 3). These challenges in data management do not affect the cost analysis in Table 2, which describes expenditures prior to the comparative study.

Safe emptying jobs performed

We confirmed that The Wastewater Association coordinated 135 emptying jobs (removing approximately 1,065,600 L of fecal sludge), Vukasasa coordinated 41 jobs (removing approximately 147,120 L of fecal sludge), and Nyalenda Water coordinated 18 jobs (removing 95,960 L of fecal sludge) during our study period (Table 3). Because our landscape assessment revealed that formal service

Table 2 | Costs of safe emptying and transport: capital expenditures (CAPEX) and operating expenditures (OPEX) incurred by pit-emptying service providers in Kisumu, Kenya

		Formal manual emptying			VTOs		
Cost category		Cost per emptying (KES)	Cost per emptying (USD)	Relative costs	Cost per emptying (KES)	Cost per emptying (USD)	Relative costs
OPEX costs	Truck rental	3,000	30	49%	–	–	–
	Salary/labor (for 1 toilet)	2,550	25.50	41%	1,117	11.17	69%
	Office costs	78	0.78	1.3%	–	–	–
	Disinfectant consumables	245	2.45	4%	–	–	–
	Maintenance	–	–	–	56	0.56	3%
	Fuel	–	–	–	232	2.32	14%
	License/membership				85	0.85	5%
Total OPEX		5,873	58.73	95%	1,489	14.89	91%
CAPEX costs	Registrations	–	–	–	8	0.08	0.5%
	Emptying equipment	77	0.77	1%	125[a]	1.251	8%1
	Safety equipment[b]	220	2.20	4%	8	0.08	0.5%
Total CAPEX		296	2.96	5%	140	1.40	9%
Total costs		6,169	61.69	100%	1,630	16.30	100%

For formal manual emptying, we report data obtained from the Vukasasa organization; costs were similar for the other two formal manual emptying groups.
Note: exchange rate is 1.00 USD = 100 KES (6 June 2019, oanda.com).
[a]For VTOs, the emptying equipment cost includes an exhauster truck and hoses; for formal manual emptiers, the truck price is captured in the truck rental fee.
[b]The amortized cost of safety equipment is higher for formal manual emptying than VTOs primarily because of economies of scale: formal emptying groups were emptying an average of five pits per month, compared to VTOs that were emptying an average of six per day. However, economies of scale would not affect the operating expenditures, which drive the costs for both services: operating costs are 95% of total formal manual emptying costs and 91% of total VTO costs.

Table 3 | Activity summary of managing groups: number of pits emptied, volume of fecal sludge emptied, and types of toilets served based on validated data

Managing group		The Wastewater Association	Vukasasa	Nyalenda Water
Number of pits emptied[a]		135	41	18
Volume of sludge emptied (L)[b]		1,065,600	147,120	95,960
Average sludge volume per job (L)		7,893	3,588	5,331
Number of pits by service provider		120 VTOs, 15 formal manual emptiers	41 formal manual emptiers (Vukasasa)	9 VTOs, 9 formal manual emptiers
Number of trips[c]		120 VTO, 39 formal manual emptier (Gasia Poa)	118 formal manual emptiers (Vukasasa)	9 VTO, 26 formal manual emptiers
Number of toilet facilities serviced[d]	Septic tank (flush/pour-flush)	13 (17%)	2 (5%)	6 (35%)
	Pit latrine (flush/pour-flush)	12 (15%)	7 (17%)	5 (29%)
	Ventilated improved pit latrine	15 (19%)	5 (12%)	1 (6%)
	Dry pit latrine with slab	38 (49%)	27 (66%)	5 (29%)

[a]We excluded all records that were not validated by independent enumerators (e.g., households not found or no emptying done) from analysis: these included 1 record for The Wastewater Association, 19 for Vukasasa, and 26 for Nyalenda Water.
[b]We estimated the volume of sludge removed by VTOs based on exhauster truck capacity. We estimated the volume of sludge removed via formal manual emptying by multiplying the number of barrels used (of either 60, 120, or 210 L capacity) with the number of trips required to transport the sludge to the treatment site.
[c]VTOs conducted one trip per pit, while the formal manual emptiers conducted an average of two to three trips per pit.
[d]Data are based on follow-up surveys: $n = 78$ for The Wastewater Association, $n = 41$ for Vukasasa, and $n = 18$ for Nyalenda water.

providers were largely inactive in the study area prior to the pilot, we can infer that the vast majority of the 1,308,680 L of fecal sludge safely emptied during this study would otherwise have been disposed of unsafely.

Formalized manual emptying required multiple trips to transport all of the fecal sludge from a single pit to KIWASCO's treatment site (generally 2–3 trips per pit): the 41 emptying jobs managed by Vukasasa required 118 trips (Table 3). The Wastewater Association subcontracted 15/135 jobs to the formal manual emptying group, Gasia Poa, which accounted for 39 trips to the treatment site (Table 3). Nyalenda Water subcontracted nine jobs to VTOs and nine to formal manual emptiers (corresponding to 26 trips to the treatment site) (Table 3). Vukasasa delegated all jobs to its members and did not subcontract any jobs to VTOs. We also found that all groups serviced a variety of toilet types (Table 3).

Financial metrics of managing groups

The cash incentives that we provided motivated all three managing groups to reduce the prices of emptying services, and, therefore, served as subsidies for their customers (Table 4). The average price charged by The Wastewater Association (1,375 KES, 14 USD per job) was substantially lower than the average prices charged by the other two managing groups (3,190 KES, 32 USD for Vukasasa and 2,696 KES, 27 USD for Nyalenda Water). These prices were comparable to those charged by illegal manual emptiers (10–30 USD) (Peletz *et al.*, forthcoming), which suggests that the three managing groups were able to compete with informal service providers.

The Wastewater Association showed a lower average net revenue (after incentive payments) per job (1,389 KES, 14 USD) compared to Vukasasa (3,124 KES, 31 USD) and Nyalenda Water (2,474 KES, 25 USD). However, the Association completed substantially more jobs and their total net revenue was higher: 187,486 KES (1,875 USD) for the Association, compared to 128,100 KES (1,281 USD) for Vukasasa and 44,527 (445 USD) for Nyalenda Water.

Customer feedback

The Wastewater Association's customers reported the highest satisfaction levels: they gave an average rating of 4.3 (on a 5-point scale) and 78% reported that they would use the

Table 4 | Financial metrics for each managing group recorded during our study period

Managing group	The Wastewater Association KES	USD	Vukasasa KES	USD	Nyalenda Water KES	USD
Average consumer payment per job	1,375	14	3,190	32	2,696	27
Average payment to service provider per job	(2,986)	(30)	(3,066)	(31)	(3,222)	(32)
Average payment to rental truck driver per job[a]	(578)	(6)	(5,756)	(58)	(2,889)	(29)
Average loss before incentive payment per job	(2,189)	(22)	(5,632)	(56)	(3,415)	(34)
Incentive payment to managing group per job	3,000	30	3,000	30	3,000	30
Average incentive payment for truck rental per job	578	6	5,756	58	2,889	29
Average managing group net revenue per job	1,389	14	3,124	31	2,474	25
Benefit-to-cost ratio	1.39		1.35		1.40	
Number of jobs	135	–	41	–	18	–
Total managing group net revenue	187,486	1,875	128,100	1,281	44,527	445

Numbers in parentheses represent losses.
Exchange rate is 1.00 USD = 100 KES (6 June 2019, oanda.com).
[a]For jobs completed by formal manual emptiers, we provided a 20 USD incentive per trip to the treatment site. The average incentive therefore is equal to the proportion of jobs completed by formal manual emptiers multiplied by the average number of trips to the treatment plant multiplied by 20 USD.

same service again at the same price (Table 5). When the Wastewater Association's customers were asked what they liked about the service, their most common response was price (56%); when asked what they did not like, only 9% responded that their pits were partially emptied and only 3% responded that the solid waste was not removed

Table 5 | Customer satisfaction across the managing groups, based on the independent follow-up survey conducted after the comparative study

		Overall (*n* = 136)	The Wastewater Association (*n* = 78)	Vukasasa (*n* = 41)	Nyalenda Water (*n* = 17)
Customer satisfaction score (5 = best, 1 = worst)		4.1	4.3	4.2	2.9
Intent to use	Yes, through this group, at the same price	70%	78%	61%	53%
	Yes, through this group, not at the same price	18%	14%	20%	35%
	No, not through this group, but at the same price	4%	3%	7%	6%
	No, not through this group, and not at the same price	4%	3%	7%	6%
	Don't know	3%	3%	5%	-
What customers liked	Price	43%	56%	17%	41%
	Quality of service	41%	40%	46%	35%
	Communication	8%	10%	5%	6%
	Proper fecal sludge removal	35%	24%	59%	24%
	Other[a]	28%	24%	30%	40%
What customers did not like	Price change after negotiation	1%	1%	0%	0%
	Pit partially emptied	15%	9%	17%	35%
	Solid waste not removed	2%	3%	2%	0%
	Smell	18%	14%	22%	24%
	Conceal from authority	18%	14%	24%	24%
	Other[b]	18%	14%	22%	24%

[a]Other includes: lack of smell, speed of emptying, and pit fully emptied.
[b]Other includes: service too expensive, sludge buried on site, sludge disposed in open.

(Table 5). Vukasasa received performance ratings comparable to the Association's (average rating of 4.2). Customers particularly appreciated the proper fecal sludge removal (59%) and the quality of service (46%) (Table 4). Nyalenda Water received the lowest average satisfaction rating of 2.9, with 35% of customers complaining that pits were not fully emptied (Table 5). Among households that did not use the subsidized service, 65% (35/54) reported they were not aware of the service, 20% (11/54) reported not needing to empty their pit, and 15% (8/54) reported that the price was too high or they lacked money; however, we did not collect information on whether households used an alternative service.

CONCLUSION

Study summary and key findings

To date, programs for improving fecal sludge management in low-income areas of Kisumu have focused on the development of formal manual emptying organizations that are recognized by local authorities, employ safety procedures and equipment, and transport fecal sludge to the KIWASCO fecal sludge treatment site. However, the respective roles of VTOs and formal manual emptiers in expanding safe sanitation services to low-income areas have not been rigorously investigated. In this study, we compared the economics of VTOs and formal manual emptying organizations. We also employed a financial incentives-based strategy to promote the expansion of safe pit-emptying services in a low-income area and compared the performance of three managing groups to coordinate these services: (1) The Wastewater Association; (2) a formal manual emptying organization; and (3) a community-based water supplier seeking to expand into pit-emptying.

Our economic analysis showed that VTOs were about four times more cost-effective than formal manual emptying organizations (Table 2). The main contributors to higher operating costs for formal manual emptying included truck rentals for transporting fecal sludge to the KIWASCO treatment site and labor expenses (Table 2). The Wastewater Association was more efficient in coordinating emptying services in low-income areas than the formal manual emptying organization and the community-based water supplier, both in the numbers of pits emptied and in the numbers of trips required to transport fecal sludge from an emptied pit to the treatment site (Table 3). The Wastewater Association also received the highest customer satisfaction ratings (Table 5) and maintained the most reliable records of business activities. These findings challenge the assumption that VTOs are not appropriate solutions for servicing low-incomes areas, at least in Kisumu, Kenya. On the contrary, VTOs can effectively service a large proportion of low-income households in Kisumu, using formal manual emptiers as a substitute in the fraction of cases where they cannot technically operate.

Service delivery model with potential to scale: sanitation 'switchboard'

The strong performance of The Wastewater Association in our comparative study was likely driven by multiple factors: their substantially lower operating costs and price points, their greater experience in marketing their services, their higher capacity for performing jobs through their membership of VTOs, and their adherence to accurate record keeping. On a pilot scale, the Association's service delivery model was successful in delivering safe sanitation: pits were safely emptied at a price point low enough to eliminate informal, unsafe pit-emptying as a viable economic substitute and households were satisfied with the Association's services. The key elements of this service delivery model were: (1) a competent local organization serving as a 'switchboard' between households and suppliers; (2) broad flexibility given to that organization to assess pits, set prices, choose appropriate suppliers, negotiate supplier payments, coordinate timing and terms, and ensure customer satisfaction by resolving disputes; (3) holding the organization accountable by careful monitoring of high-level outcomes and customer satisfaction; (4) providing an outcome-based incentive.

Our study clearly demonstrated that the choice of the organization matters. The organization needs to be visible and trusted by the community but also quite sophisticated in its business practices; experience in sanitation service provision is also critical. Another key element of the model is to entrust operational decisions to the organization, rather than be prescriptive from the top down (for process, such as selecting suppliers, setting prices, marketing activities, etc.). For example, The Wastewater Association outsourced 11% (15/135) of their emptying jobs to the formal manual emptying group, Gasia Poa, because vacuum trucks could not reach the pits, or because the pits contained solid waste that required manual removal, or a combination of both factors (Table 3). This 'hybrid' service delivery approach combines the strengths of the two types of service providers and potentially promotes investments by the VTOs in the manual emptying organizations. The 'switchboard' model has substantial potential to scale in Kisumu (with the Wastewater Association) and beyond. Comparable 'switchboard' models in Dakar, Senegal, and Dhaka, Bangladesh, have shown promising results in expanding safe pit-emptying services in low-income areas, with the municipal utility and/or a private entity acting as the 'managing group' (ONAS 2014; FSM4 2017). The most appropriate organization to operate as a switchboard on a large scale will likely vary from city to city; the selection, training, and monitoring of this organization require careful consideration.

Addressing affordability

Service providers applied our payment incentives to subsidize their emptying fees: for example, despite estimated costs of 6,169 KES (62 USD) for formal manual pit-emptying, Vukasasa charged customers an average of 3,190 KES (32 USD) during our comparative study (Tables 2 and 4). These subsidies likely expanded their markets during the study period, as poor households are typically unable to afford safe pit-emptying services (Burt Sklar & Murray 2019; Peletz et al., forthcoming). Our study thus validated that financial incentives are effective at promoting the expansion of formal pit-emptying services in low-income areas. Identifying the optimal incentive amount to support regulated pit-emptying service delivery models at scale will require accurate measurements of WTP for pit-emptying among poor households and careful consideration of the prices charged by informal manual pit emptiers (J-PAL 2012; Daudey 2017). Further, identifying sustainable resources for these subsidies, whether from donor funds, government budgets, pro-poor sanitation surcharges imposed on utility bills, or higher fees charged to the wealthy (cross-subsidies), is an essential political consideration, if Kisumu is committed to improved fecal sludge management (Blackett & Hawkins 2017; Acey *et al.* 2019). For example, the cities of Wai and Sinnar in India finance inclusive scheduled desludging services through a surcharge on property taxes (Mehta Mehta & Yadav 2019).

Expanding treatment capacity and regulation

As a final political consideration, we emphasize the importance of increasing Kisumu's capacity and regulation of fecal waste treatment, both for disposal and reuse. As noted in the introduction, estimates indicate that the KIWASCO treatment site only receives about one-third of the fecal sludge generated in the city (Furlong & Jooust 2016). Programs to increase safe pit-emptying services will likely require augmentation of Kisumu's fecal sludge treatment and disposal infrastructure, including additional staff, drying fields, and standards for chemical use (e.g., odor neutralizers) during pit-emptying. The development of the Kisumu County Environmental Sanitation and Hygiene Policy and the Kisumu County Environmental Health and Sanitation Bill are promising steps for increased regulation for fecal sludge treatment (County Government of Kisumu 2018a, 2018b).

ACKNOWLEDGEMENTS

This research was delivered through a research agreement between Water & Sanitation for the Urban Poor (WSUP) and The Aquaya Institute. The agreement was funded under the Urban Sanitation Research Initiative managed by WSUP and supported by UK Aid from the UK government; we appreciate the contributions from the WSUP team including Guy Norman, Sam Drabble, Rosie Renouf, and Emanuel Owako. We gratefully acknowledge our enumerators, interviewees, managing groups, FGD participants, households, and working committee members that contributed to this study, especially Jane Mumma and Sheillah Simiyu from Great Lakes University Kisumu. We also thank Aquaya staff members, Alicea Cock-Esteb, Edinah Samuel, and Brian Mwangi, for their assistance with data collection and management.

DATA AVAILABILITY STATEMENT

All relevant data are included in the paper or its Supplementary Information.

REFERENCES

Acey, C., Kisiangani, J., Ronoh, P., Delaire, C., Makena, E., Norman, G., Levine, D., Khush, R. & Peletz, R. 2019 Cross-subsidies for improved sanitation in low-income settlements: assessing the willingness to pay of water utility customers in Kenyan Cities. *World Development* **115**, 160–177. https://doi.org/10.1016/j.worlddev.2018.11.006.

Balasubramanya, S., Evans, B., Hardy, R., Ahmed, R., Habib, A., Asad, N., Rahman, M., Hasan, M., Dey, D., Fletcher, L., Camargo-Valero, M., Rao, K. & Fernando, S. 2017 Towards sustainable sanitation management: establishing the costs and willingness to pay for emptying and transporting sludge in rural districts with high rates of access to latrines. *PLoS ONE* **12** (3), 1–20. https://doi.org/10.1371/journal.pone.0171735.

Blackett, I. & Hawkins, P. 2017 *Case Studies on the Business, Policy and Technology of Faecal Sludge Management*, 2nd ed. Bill & Melinda Gates Foundation, Seattle, WA.

Burt, Z., Sklar, R. & Murray, A. 2019 Costs and willingness to pay for pit latrine emptying services in Kigali, Rwanda. *International Journal of Environmental Research and Public Health* **16** (23), 4738. https://doi.org/10.3390/ijerph16234738.

Castro, V. 2009 *Improving Water Utility Services Through Delegated Management: Lessons From the Utility and Small-Scale Providers in Kisumu, Kenya*. Nairobi, Kenya. Available from https://www.wsp.org/sites/wsp.org/files/publications/Af-imp_through_delegated_mgmt.pdf

Chowdhury, S. & Kone, D. 2012 *Business Analysis of Fecal Sludge Management: Emptying and Transportation Services in Africa and Asia.*

County Government of Kisumu 2018a *Kisumu County Annual Development Plan (FY 2018/2019).*

County Government of Kisumu 2018b *Kisumu County Environmental Sanitation and Hygiene Policy 2030 (Draft).*

Daudey, L. 2017 The cost of urban sanitation solutions: a literature review. *Journal of Water, Sanitation and Hygiene for Development* **8** (2), 176–195. https://doi.org/10.2166/washdev.2017.058.

FSM4 2017 *FSM4 Case Studies*. Chennai, India.

Furlong, C. & Jooust, L. 2016 *SFD Report Kisumu, Kenya.*

J-PAL 2012 *Improving Access to Urban Services for the Poor: Open Issues and A Framework for A Future Research Agenda.* Abdul Latif Jameel Poverty Action Lab, Cambridge, MA.

Mansour, G., Oyaya, C. & Owor, M. 2017 *Situation Analysis of the Urban Sanitation Sector in Kenya.*

Mehta, M., Mehta, D. & Yadav, U. 2019 Citywide inclusive sanitation through scheduled desludging services: emerging experience from India. *Frontiers in Environmental Science* **7**, 188. https://doi.org/10.3389/fenvs.2019.00188.

Nzengya, D. M. 2015 Exploring the challenges and opportunities for master operators and water kiosks under delegated management model (DMM): a study in Lake Victoria region, Kenya. *Cities* **46**, 35–43. https://doi.org/10.1016/j.cities.2015.04.005.

ONAS 2014 Fecal sludge management program: Lessons learned. *Quarterly Magazine of the ONAS FSM Program.*

Owako, E. & Renouf, R. 2018 *Strengthening the Enabling Environment for Urban Sanitation: Public–Private Collaboration in Kisumu, Kenya.*

Peletz, R., MacLeod, C., Kones, J., Samuel, E., Easthope-Frazer, A., Delaire, C. & Khush, R. (in preparation) When pits fill up: supply and demand for safe pit-emptying services in Kisumu, Kenya.

Practical Action 2016 *KisumuSan Baseline Survey Final Report.* Kisumu, Kenya.

Schwartz, K. & Sanga, A. 2015 Partnerships between utilities and small-scale providers: delegated management in Kisumu, Kenya. *Physics and Chemistry of the Earth* **35** (13–14), 765–771. https://doi.org/10.1016/j.pce.2010.07.003.

WSUP 2018 *Regulatory Reform: Raising the Standard of pit-Emptying Services in Kisumu.*

First received 17 March 2020; accepted in revised form 3 July 2020. Available online 5 August 2020

Research Paper

Users are willing to pay for sanitation, but not as much as they say: empirical results and methodological comparisons of willingness to pay for peri-urban sanitation in Lusaka, Zambia using contingent valuation, discrete choice experiments, and hedonic pricing

James B. Tidwell

ABSTRACT

Significant investment is needed to improve peri-urban sanitation. Consumer willingness to pay may bridge some of this gap. While contingent valuation has been frequently used to assess this demand, there are few comparative studies to validate this method for water and sanitation. We use contingent valuation to estimate demand for flushing toilets, solid doors, and inside and outside locks on doors and compare this with results from hedonic pricing and discrete choice experiments. We collected data for a randomized, controlled trial in peri-urban Lusaka, Zambia in 2017. Tenants were randomly allocated to discrete choice experiments ($n = 432$) or contingent valuation ($n = 458$). Estimates using contingent valuation were lower than discrete choice experiments for solid doors (US$2.6 vs. US$3.4), higher for flushing toilets ($3.4 vs. $2.2), and were of the opposite sign for inside and outside locks ($1.6 vs. $ − 1.1). Hedonic pricing aligned more closely to discrete choice experiments for flushing toilets ($1.7) and locks (−$0.9), suggesting significant and inconsistent bias in contingent valuation estimates. While these results provide strong evidence of consumer willingness to pay for sanitation, researchers and policymakers should carefully consider demand assessment methods due to the inconsistent, but often inflated bias of contingent valuation.

Key words | contingent valuation, discrete choice experiments, financing, hedonic pricing, sanitation, willingness to pay

James B. Tidwell
Mossavar-Rahmani Center for Business and Government,
Harvard Kennedy School of Government,
Cambridge, MA,
USA
and
World Vision,
Washington, DC,
USA
E-mail: *btidwell@worldvision.org*

HIGHLIGHTS

- A commonly used method for assessing willingness to pay, contingent valuation, produces inconsistently biased and often inflated values.
- While there is significant willingness to pay for some aspects of sanitation by users, methods that elicit preferences indirectly and/or through real market transactions should be used where possible to inform policymakers and the private sector.

INTRODUCTION

People who live in peri-urban areas are a large and growing proportion of the population, and the challenges of ensuring adequate sanitation in these areas are a significant public health challenge that requires understanding and leveraging household investment to reach the sustainable development goal of 'Sanitation for All'. Peri-urban areas suffer from poor

doi: 10.2166/washdev.2020.072

quality infrastructure and health outcomes are worse in these areas than either urban or rural areas (Ezeh *et al.* 2017). The number of people living in peri-urban areas is increasing rapidly around the world and is estimated to reach 2 billion people by 2035 (UN-HABITAT 2003).

Inadequate sanitation is a major cause of lost DALYs (GBD Diarrhoeal Diseases Collaborators 2017) and results in more than $220 billion USD of economic losses annually (Lixil, WaterAid, and Oxford Economics 2016). The cost of reaching the SDG (Sustainable Development Goals) for sanitation is estimated at more than $1 trillion USD during the SDG era (Hutton & Varughese 2016). The anticipated financing needs are greatest for urban and peri-urban areas, and the burden is expected to fall largely on governments and donors with little consideration of the role of the substantial direct household investment required of households and indeed already taking place (Danert & Hutton 2020). Along with ensuring the security of land tenure (Scott *et al.* 2013), advocating for high-quality shared sanitation to be included in the SDGs (Evans *et al.* 2017), and providing ways to increase willingness to pay through marketing messages or credit products (Ben Yishay *et al.* 2017), estimating and leveraging this demand is key to ensuring adequate sanitation for people living in peri-urban areas.

We present here the results of a study of tenant willingness to pay for higher-quality shared sanitation in peri-urban Lusaka, Zambia. Peri-urban areas are rapidly growing and present significant health and sanitation challenges globally (Nakagiri *et al.* 2016; Ezeh *et al.* 2017), and a lack of hygienic conditions, desirability, accessibility, and sustainability (Tidwell *et al.* 2018a) lead to a number of negative health and well-being impacts (Shiras *et al.* 2018). Sanitation is often not directly purchased in the market in these settings, but instead is selected along with a choice of housing in one bundled rental payment, making willingness to pay assessment more difficult. We collected data from 1,085 landlords and their tenants using both stated and revealed preference methods to estimate tenant willingness to pay and implications for sanitation promotion and service delivery in this setting.

Understanding user preferences and willingness to pay is generally done using either stated preferences (SPs), hypothetical choices expressed by users, or revealed preferences (RPs), which come through observing decisions with real-life stakes in existing markets or experimental settings. Stated preference methods can be either direct, asking a respondent what they are willing to pay for something, or indirect, such as asking a respondent to rank or choose between alternatives, which can be used to calculate implicit willingness to pay. Stated preference methods are frequently used because they are easier and cheaper to administer and can be used to value products or services that do not yet exist. The contingent valuation method (CVM) (Mitchell & Carson 1989) is an SP method that directly elicits whether a consumer is willing to pay a proposed amount. Though CVM was originally developed for understanding preferences for environmental quality, where there was no observable market (Hanemann 1994), it has often been applied to exploring potential markets for water and sanitation products (Whittington *et al.* 1993; Altaf & Hughes 1994; Null *et al.* 2012; Sehreen *et al.* 2019). Indirect methods such as discrete choice experiments (DCEs) are being used (Quaife *et al.* 2018), where respondents are presented a series of choices between alternatives with different characteristics, which may avoid some of the potential biases of direct SP methods.

The most significant concern with using SP methods is hypothetical bias (Hausman 2012), which occurs when answers differ between hypothetical scenarios and real-life scenarios when respondents must make real money transactions. Hypothetical bias is widely recognized in the SP literature (List & Gallet 2001; Little & Berrens 2004; Murphy *et al.* 2005), and a recent meta-analysis found that CVM inflated mean WTP (Willingness to Pay) estimates by a factor of 1.79. However, CVM studies have been conducted most often in situations where consumers have no market experience (such as for air quality or presence of recreational parks), and there is less evidence about hypothetical bias for CVM in settings where consumers have purchasing experience (Hanemann 1994; Carson *et al.* 2001).

There is less evidence about the magnitude of hypothetical bias in DCEs, with some finding higher marginal WTP from SP (Johansson-Stenman & Svedsäter 2003), others suggesting they are equal (Carlsson & Martinsson 2001; Cameron *et al.* 2002; Tidwell *et al.* 2019b), and one even finding that DCEs produced lower WTP estimates due to routine behaviors resulting in increased RP values (Isacsson 2007). In addition, DCEs often predict actual health behaviors (Quaife *et al.* 2018), especially when differences in personal preferences are taken into account in the models used (de Bekker-Grob *et al.* 2020), and that they are less subject to social desirability bias than other SP methods (Horiuchi *et al.* 2018).

Revealed preference methods aim to overcome problems of hypothetical bias by requiring monetary stakes to be associated with decision-making. The price of sanitation associated with a rental decision cannot be separately observed in the market, but by using an RP method called the hedonic pricing method (HPM), we can calculate the implicit price of each housing and sanitation attribute (Rosen 1974). However, RPs may suffer from biases and empirical challenges as well. They may be biased if choices are complex (Beshears *et al.* 2008) or when there is market failure (Hanna & Richards 2014). HPM requires pre-selecting the model used to estimate willingness to pay (van den Berg & Nauges 2012) and rare combinations of attributes and collinearity make model estimation challenging (Harrison & Rubinfeld 1978). HPM has been applied to sanitation in a few cases where a large fraction of rent was being estimated, such as for the presence of a toilet (Simiyu *et al.* 2017) or a sewer connection (Vásquez & Beaudin 2020).

SP studies, almost exclusively using CVM, have found significant willingness to pay for sanitation, including urban residents willing to pay 14% of mean monthly expenditure for high-quality on-site sanitation (Altaf & Hughes 1994) and rural households willing to pay almost a third of their annual income for a flushing toilet (Van Minh *et al.* 2013). However, RP studies have produced much smaller estimates of willingness to pay. For example, using voucher-based methods, less than 5% of a rural Tanzanian population was willing to pay even half the market price of improved latrine slabs, where the full price amounted to less than 5% of their annual income (Peletz *et al.* 2017).

Reliable WTP estimates are important for policymakers and may be used to understand the market potential of new products (Van Minh *et al.* 2013) or optimal government subsidies to drive improvements (Whittington *et al.* 1993). While a previous study using this dataset focused on the DCE results in comparison to hedonic pricing values (Tidwell *et al.* 2019b), DCEs are more complicated to design and analyze than CVM studies, and so, CVM studies would be preferable if shown to produce reliable estimates. Therefore, in this study, we report the results the CVM approach to assessing willingness to pay for improvements in peri-urban shared sanitation quality compared to these other methods. We further suggest the reasons for differences between the approaches and implications for method selection and interpretation for policymakers.

METHODS

Study setting

The data for this study were collected as part of a trial of a behavior-change intervention targeting landlords in Bauleni Compound, a peri-urban neighborhood in Lusaka, Zambia. Bauleni has an approximate population of 64,000 people, and it is subdivided into plots of land, which were originally intended for one household. However, landowners have become *de facto* landlords, with a median of three tenant households [IQR: 2–4] living on each plot at present (Tidwell *et al.* 2019b). We surveyed 1,137 landlords, and after excluding 52 who had no current tenants, enrolled 1,085 landlords in our trial. For each landlord enrolled, data were also collected from one randomly selected adult from a tenant household living on the plot. More data about the study setting are available elsewhere (Tidwell *et al.* 2018b). There are no public toilets in the neighborhood other than in a central market, and all plots enrolled in the study had at least one toilet on the plot that could be freely accessed by the tenants, who never paid for this access other than as bundled in the monthly rent payment. Our assessment of sanitation quality in the area showed that 87% of toilets were 'improved, but shared', in terms of JMP classification, but often were unhygienic and structurally poor (Tidwell *et al.* 2018a). Therefore, we focused on understanding how much variation in rent was attributable to differences in sanitation quality.

Data collection and analysis

The data for this study were collected in June and July of 2017 from a total of 1,085 landlord-tenant pairs. Hedonic pricing was based on observations of the housing and toilet characteristics, while stated preference methods were used to collect WTP data directly from tenants. Locally based enumerators were trained by our local research partner, the Center for Infectious Disease Research in Zambia (CIDRZ). Data were collected using a systematic random sample of every fourth house within Bauleni. We conducted a round of pilot data collection in a nearby neighborhood to ensure high-quality data were being collected and to generate priors from which to design the DCEs. Data were collected using ODK Collect software (Hartung *et al.* 2010), and analysis was done using R version 3.5.2 (R Core Team 2018).

The HPM used a regression analysis with the household's monthly rental payment as the dependent variables and a model that incorporates their housing attributes (including house, plot, and community-level factors). Analysis was conducted with a generalized linear model with the first model incorporating sanitation through a simple binary variable capturing the presence of a toilet and a second model that measured sanitation quality through a number of more granular measures (such as the presence of a solid door and the presence of a concrete slab). The models consisted of a standard linear combination of attributes (Champ *et al.* 2017), other than an interaction term between the presence of water on the plot and a flushing toilet in the second model, which emerged as interrelated considerations in the formative research stage (Tidwell *et al.* 2018b). For the one case where a measured attribute had another as prerequisite (i.e., a solid door is required to have inside and outside locks), we modeled this is a categorical variable and performed a *post hoc* analysis to separate out the value of inside and outside locks. We estimated both models using OLS (Ordinary Least Squares) and tested for heteroscedasticity, variance inflation due to multicollinearity, and functional form.

DCEs presented a series of six pairs of choices between rental accommodates, where housing characteristics were fixed, but with different kinds of toilets. Toilet options varied by the presence of a simple cover (not included in the contingent valuation questions, and thus not included here), flushing toilet, solid door, inside and outside locks, and price. Tenants then selected the rental accommodation that they preferred based on the toilet options presented. Additionally, tenants were randomly assigned to one of two separate groups of choice sets that differed based on whether a water source was present on the plot, which was suggested to be an important consideration based on feedback during our piloting. DCEs were analyzed using four different logit models including fixed and random parameters and with and without price interactions, and ultimately, a random parameters model with price interaction was selected based on having the lowest AIC3 value (Andrews & Currim 2003). This analysis produced sample-level estimates of mean and standard deviation for WTP for each of the included attributes, which were calculated using the ratio of parameter estimates for each attribute to the estimate for the price attribute. More details about the HPM and DCE analysis are available elsewhere (Tidwell *et al.* 2019b).

Contingent valuation

A series of dichotomous choice contingent valuation questions were asked of each respondent to solicit their WTP for several toilet improvements based on standard guidance (Boyle 2017). We asked about upgrading from a simple hole to a flushing toilet (on a plot with a water tap – see discussion below), from a toilet without a solid door to one with a solid door (without locks on the inside or outside), and from a solid door without any locks to a solid door with both locks. These upgrades were presented as generic (unbranded) upgrades, with a generic ceramic flushing toilet seat, a basic, unfinished solid wood door, and a standard mount for a padlock (outside lock) and deadbolt lock (inside lock) shown/described. Inside and outside locks were grouped together as the anticipated willingness to pay for either was small, due to their low cost. Each of these upgrades was found in a substantial number of existing toilets in the area, but there is not a large variety of types available, and thus, heterogeneity of WTP due to differing assumptions of product quality was considered to be minimal.

Tenant household members identified as being primarily responsible for financial decision-making regarding housing choices were asked if they would be willing to pay more rent if they were hypothetically living on a plot that did not already have the improvement, but the landlord made each of three improvements. If they said yes, they were asked if they would pay a certain amount more rent if a landlord made the improvement, with a randomly assigned monthly rent increase value of 10–50 Kw in 10 Kw (US$1) increments (for the flushing toilet and solid door) and 5–25 Kw in 5 Kw (US$0.5) increments (for adding locks). Tenants were reminded to consider their real-life budget constraints to reduce the possibility of hypothetical bias.

Utility was modeled using a standard approach (Hanemann & Kanninen 2001), where a combination of observed and unobserved variables are included for each individual and maximum likelihood estimation is used to estimate the parameters of the utility function and the variance of the normally distributed error term, and these values are then used to calculate WTP. Analysis was conducted using the *sbchoice* package in R, with truncated mean WTP estimated using a normal distribution (Estimation using a logistic distribution yielded similar results, and log-logistic, log-normal, and Weibull distributions performed poorly on the dataset because of the large number of positive responses to questions, even at the high end of the bid price range) and standard errors calculated using bootstrapping, except where low WTP led to errors in the bootstrapping procedure. In these cases, the Krinsky–Robb method for calculating confidence intervals (Krinsky & Robb 1986) was used instead, though it produces more conservative confidence intervals (Hole 2007). Tenant income, gender, number of household members, age, level of education, and employment type were included in the model as covariates.

Ethical considerations

Enumerators received training on following ethical procedures of data collection. They delivered the surveys in person to landlords and tenants after reading aloud an information sheet in English or a local language (Bemba or Nyanja), responding to any questions, and receiving written consent or a thumbprint with witness present if the person was unable to write their name. Participants were informed that they were allowed to terminate the survey at any time for any reason. No compensation was provided to any participant, and no personally identifying information was collected. Both the University of Zambia Biomedical Research Ethics Committee in Lusaka, Zambia (ref: 002-02-17) and the London School of Hygiene and Tropical Medicine (ref: 12157) approved the study.

RESULTS

Sample characteristics

A total of 890 tenants were surveyed with one of the two willingness to pay methods out of the 1,085 landlords enrolled (Table 1), with fewer tenants than landlords surveyed due to challenges in scheduling a suitable follow-up time for an interview (if the selected tenant was not present

Table 1 | Sample demographic characteristics by willingness to pay method

Characteristic	CVM	DCE	*p*-value
Number of observations	458	432	
Gender = female	80.8%	81.9%	0.721
Age (mean (SD))	30.0 (9.0)	30.8 (9.9)	0.184
Monthly rent (Kw) (mean (SD))	481 (192)	469 (198)	0.369
Monthly household income (Kw) (mean (SD))	1,318 (1,107)	1,320 (1,071)	0.979
Number of household members (mean (SD))	3.1 (1.8)	3.2 (1.9)	0.298
Respondent education level			
Primary or less	20.5%	20.1%	0.874
Some or completed secondary	72.3%	74.5%	0.447
Beyond secondary	7.2%	5.4%	0.263
Respondent employment type			
Unemployed/housework	50.9%	48.4%	0.458
Casual worker/piece work	13.3%	13.0%	0.875
Self-employed/business	22.1%	26.4%	0.131
Formal employee	10.9%	9.0%	0.348
Retired	0.2%	0.2%	0.967
Other	2.6%	3.0%	0.726
Marital status			
Married	73.1%	73.8%	0.814
Single, never married	18.1%	16.0%	0.395
Divorced/separated	8.7%	10.0%	0.532

Note: Differences calculated between individual responses within categorical variables.

at the time of the landlord interview) or refusal by the selected tenant. Roughly equal numbers were assigned to CVM ($n = 458$) and DCE ($n = 432$) arms by random assignment within the mobile data collection software package. Demographic characteristics were well balanced between arms allocated to the different data collection methods. The mean age was 30 years old, and a majority of respondents were female (81.3%). Women were found in our formative research to be more influential in household decision-making related to sanitation quality, likely because of the larger amount of time they spent at home (Tidwell *et al.* 2018b). Due to the lack of gender balance in the sample, with more than 80% of respondents being female, each analysis was disaggregated by gender. About half (48.5%) of tenants earned less than US$1.25 per household member per day (Median: US$1.35, IQR: [0.79, 2.47]). Tenants paid about 36% of their monthly income for rent, and a majority had completed some or all of secondary school.

Logistic regression models of the utility associated with each sanitation improvement by regressing respondent characteristics and bid amounts on stated willingness to pay (a binary, yes/no variable) showed little variation across covariates included in each model (Table 2). The level of income was negatively associated with absolute WTP for doors, but not for other improvements at the $p < 0.05$ level. Bid amounts were only related to rate of yes responses for flushing toilets at the $p < 0.05$ level, though overall rates of responding yes to CVM questions were low for the other two improvements. No geographic component was included in the analysis as the survey took place in one neighborhood and zones created artificially for the survey collection process showed no differences across housing covariates in earlier analyses (Tidwell *et al.* 2019b).

Table 2 | Analysis of covariates of willingness to pay for sanitation improvements

Outcome = 1 if willing to pay for toilet component at offered bid value:	**Flushing toilet (Estimate (Std Err))**	**Solid door (Estimate (Std Err))**	**Inside and outside locks (Estimate (Std Err))**
(Intercept)	0.359 (0.314)	−0.173 (0.321)	−0.399 (0.337)
Income (thousand Kw)	−0.011 (0.066)	−0.141 (0.071)**	−0.132 (0.072)*
Age	0.004 (0.008)	−0.001 (0.008)	−0.001 (0.008)
Sex = Male	−0.071 (0.185)	−0.126 (0.191)	−0.294 (0.196)
Education (ref: primary or less)			
Some or completed secondary	0.07 (0.161)	0.055 (0.166)	−0.13 (0.165)
Beyond secondary	0.211 (0.305)	0.149 (0.317)	−0.087 (0.316)
Employment (ref: unemployed/houseworker)			
Casual worker	0.017 (0.201)	0.09 (0.21)	−0.037 (0.214)
Self-employed/business	−0.013 (0.161)	0.198 (0.166)	0.106 (0.167)
Formal employee	0.352 (0.235)	0.335 (0.232)	0.399 (0.235)
Retired	4.803 (146.9)	−4.132 (146.9)	5.433 (146.9)
Other	−0.582 (0.408)	−0.443 (0.471)	−0.141 (0.44)
Marital status (ref: married)			
Single, never married	0.043 (0.18)	−0.132 (0.185)	0.021 (0.185)
Divorced/separated	−0.119 (0.233)	−0.256 (0.245)	−0.39 (0.256)
Bid amount	−0.009 (0.004)**	−0.005 (0.004)	0.015 (0.009)*
N	451	451	451
Log-likelihood	−224.24	−169.12	−108.84
Pseudo-R^2	0.2618	0.4029	0.6115

Notes: Parameter estimates from logistic regression models were used to assess individual characteristics associated with willingness to pay.
* $p < 0.10$, ** $p < 0.05$, *** $p < 0.01$.

Though overall 'yes' response rates were only slightly decreasing for bids within each offered upgrade, differences between 'yes' responses for any willingness to pay and offered bids and across categories varied significantly (Figure 1). For flushing toilets, 68.7% said they were willing to pay some amount, but the number of yes responses declined to 40.1% for solid doors and 34.4% for inside and outside locks.

Willingness to pay for each of the three sanitation improvement categories was positive and statistically significant for CVM (Table 3). Mean estimates of 33.5 Kw for flushing toilets, 26.1 Kw for solid doors, and 15.6 Kw for inside and outside locks represented 7.4, 5.8, and 3.5% of median monthly rent (450 Kw [IQR: 350–550]) in the sample, which was just under half of median tenant monthly income of 1,000 Kw [IQR: 750–1,700]. All estimates were statistically significantly different from zero, and parameter estimates were all considerably larger than those from the same improvements from HPM. WTP for adding locks to a solid door was also positive, in contrast to the results of HPM. Estimates did not vary significantly based on respondent income. A sub-group analysis of respondents with zero vs. non-zero WTP for each attribute (Supplementary Table S1) found that there were no statistically significant differences by observable characteristics.

As previously reported, WTP did not differ significantly between DCEs and HPM for flushing toilet or for inside and outside locks, though WTP for a solid door was higher in DCEs than in HPM. This may have been due to differences in how surveyors assessed solid doors (fully attached with no holes) compared to how tenants perceived them (anything that could be moved to fully block the entrance for privacy) as well as collinearity between solid doors and other infrastructure components for HPM (though tests for

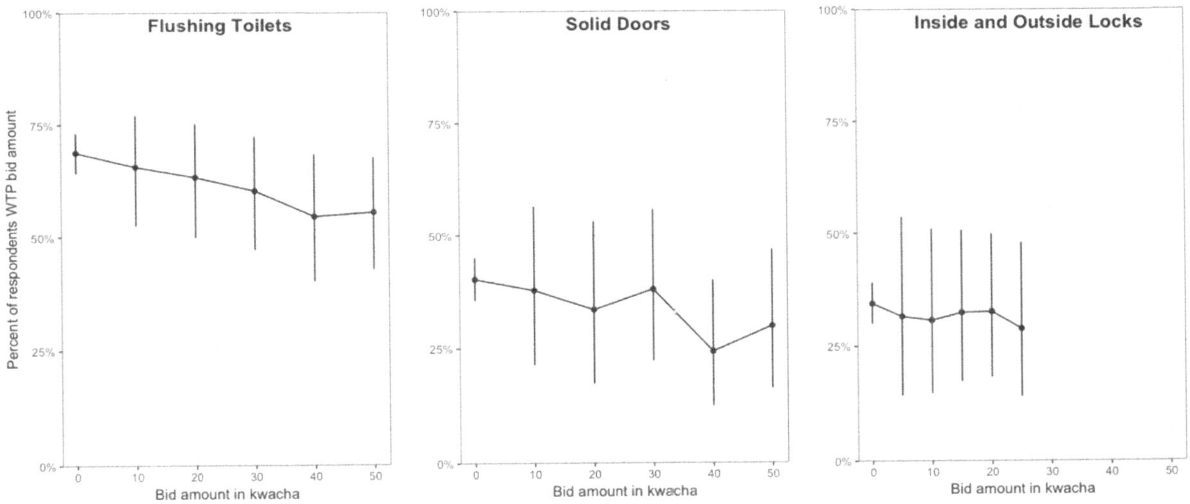

Figure 1 | Responses to contingent valuation questions by toilet improvement. *Note:* Proportion of respondents stating they were willing to pay for initial bid amount (zero) and randomly selected higher amount in 5 or 10 Kw increments for those responding yes to the initial bid; 10 Kw = US$1.

Table 3 | WTP estimates for sanitation components across WTP methods

	Contingent valuation	**DCEs**	**Hedonic pricing**			
Component	**WTP estimate in Kwacha [95% confidence interval]**			**CVM – DCE**	**CVM – HPM**	**DCE – HPM**
Flushing toilet	33.5 [30.7, 35.7]	22.1 [16.9, 27.2]	16.5 [−12.3, 45.3]	11.4 [5.6, 17.2]*	17.0 [−12.0, 46.0]	5.6 [−34.9, 23.7]
Solid door	26.1 [21.7, 28.8]	33.8 [30.2, 37.3]	3.5 [−16.5, 23.5]	−7.7 [−13.3, −2.1]*	22.6 [2.1, 43.0]*	30.3 [10.0, 50.6]*
Inside and outside locks	15.6 [9.7, 20.9]	−10.7 [−16.3, −5.1]	−8.8 [−29.0, 11.3]	26.3 [18.1, 34.4]*	24.4 [3.4, 45.4]*	−1.9 [−22.8, 19.1]

Note: Values reported in Kwacha: 10 Kw = US$1.
* $p < 0.05$.

inflation due to multicollinearity did not reject the null, aspects of toilet superstructure such as walls, doors, and a roof had VIFs of about 1.3, higher than other model parameters). For flushing toilets, CVM estimates were about 1.5 times higher than DCEs and about twice as high as HPM. For solid doors, CVM estimates were slightly lower than DCE estimates and, though the null hypothesis that CVM and HPM produced different estimates was not rejected at the $p = 0.05$ level, there was large uncertainty in the HPM estimate, and the mean estimated difference was large. For inside and outside locks, CVM estimates were much higher than DCE and HPM estimates. Notably, while estimates for the presence of inside and outside locks were positive for CVM (15.6 Kw), they were negative for both DCEs (−10.7 Kw) and for HPM (−8.8 Kw).

DISCUSSION

Estimates of WTP using the direct stated preference CVM were similar to estimates from both SPs using the indirect DCE approach and RPs estimated using the HPM for solid doors. However, CVM was much higher for flushing toilets as well as inside and outside door locks. DCEs and HPM produced generally similar WTP estimates. Poorly specified HPM parameter estimates were a challenge for each toilet characteristic examined, as there was no evidence of non-zero WTP for any particular aspect of sanitation. Standard errors were much smaller for both kinds of stated preference WTP estimates (CVM and DCE) than for HPM, which is generally the case due to the ability to control the design of stated preference methods. Hedonic pricing, based on a regression analysis and observed toilet characteristics existing in the study setting, suffered from collinearity and low prevalence of certain toilet components in HPM, leading to poorer parameter specification.

WTP generally drops off rapidly for positive prices for preventive health products and services (Cohen 2019), but the results for CVM were either level or slightly decreasing in this study for each of the examined sanitation improvements. This may be for at least two reasons. First, the range of offered prices was low relative to mean income in this population, though the prices were in line with what was seen in the market. This may have also interacted with a second factor, the likely social desirability bias for stated willingness to pay for sanitation improvements, as we saw no change in responses even by income level. CVM may be particularly unreliable for small prices such as the amortization of an improvement through a monthly rental payment. Certainly, more study is needed on this point.

It is common in the literature to assume that while CVM may overstate WTP, that it is enough to calculate an overall inflation factor by which to proportionately reduce all estimates, though satisfactory ways to do this have not been developed (Murphy *et al.* 1998; Hensher 2010; Schmidt & Bijmolt 2019). This study provides evidence that this is not sufficient for two reasons. First, CVM displayed considerable hypothetical bias for some toilet characteristics and not others. While estimates for flushing toilets were much higher via CVM than other methods, estimates for solid doors were more comparable to other methods, at least in terms of absolute differences. We hypothesize that this is due to a social desirability bias that might exist for reporting to a researcher a desire for a flushing toilet and not for solid doors, given that there is evidence that choice experiments and other indirect preference elicitation methods are less prone to social desirability bias (Horiuchi *et al.* 2018). This bias is likely inherent to a direct elicitation stated preference method like CVM.

Second, CVM may result not just in inflated estimates of WTP, but in estimates with the wrong sign. This cannot be accounted for by merely suggesting that CVM produces inflated estimates or by suggesting a proportion by which WTP estimates should be reduced to make them better reflect real demand. This issue is not inherent to CVM, but rather results from a problem in how the standard approach applies in this particular situation, which suggests at the very least that CVM may need to be augmented by a question beyond the standard initial question ('Would you be willing to pay any amount for this item?') to also include 'Would you accept it for free?' or 'Would you prefer to have or not to have this item?' or even include negative offers (i.e., willingness to accept this state in exchange for payment). This may especially be true where a product's costs outweigh its benefits for some in a heterogeneous population.

Given these inconstant biases, we should not have grouped the inside and outside locks in the design of our stated preference methods. While initially done to ensure that willingness to pay would be large enough to be detected from our sample size, it may have obscured a small positive willingness to pay for an inside lock with a larger negative willingness to pay for an outside lock (based on feedback from program participants). This also demonstrates that willingness to pay may be more related to user experience, rather than the actual value of the improvement made.

Over the last several decades, CVM studies argued that tenants were willing to pay 14% of mean monthly expenditure in urban Burkina Faso for high-quality on-site sanitation (Altaf & Hughes 1994), that rural households in Vietnam will pay almost a third of their annual income for a flushing toilet (Van Minh *et al.* 2013), that households in Delhi, India (Chopra & Das 2019) and Dhaka, Bangladesh have high willingness to pay for wastewater treatment services (Sehreen *et al.* 2019), and that wealthier residents of two Kenyan cities were willing to pay a cross-subsidy on their water bill to provide financing to improve sanitation services for poor customers resulting in $200,000 and $470,000 USD in financing annually (Acey *et al.* 2019). The results of this study suggest that such estimates are likely inflated, and may be so in a way that is difficult to simply 'adjust for' using some kind of assumption about a fixed inflation factor.

HPM has been used to assess the amount of rent attributable to differences in sanitation quality, with the presence of a toilet resulting in rent increases from 1.6% (Gulyani *et al.* 2012) to 18% (Tidwell *et al.* 2019b) to 60% (van den Berg & Nauges 2012). Increases of 16% from moving from a pit latrine to a flush toilet (Knight *et al.* 2004), 14% between shared and private toilets (Brueckner 2013), and 15% for the presence of solid walls and a roof (Tidwell *et al.* 2019b) have also been observed. However, the challenge of pre-specifying a model of willingness to pay means that these values are subject to the relatively unrealistic assumption that value consists of a linear combination of attribute values, and thus, using additional methods to validate HPM results may be warranted. DCEs may be more relevant for judging relative valuations of different attributes than CVM, given its inconsistent biases.

Toilets shared by several tenant households may be improved to some degree by leveraging tenant WTP. However, since willingness and ability to pay for high-quality sanitation are less than its costs for many, policy-relevant and practical approaches are urgently needed to address this gap, including lower-cost or subsidized products, demand creation (Tidwell *et al.* 2019a) or regulatory enforcement (Antwi-Agyei *et al* 2019) interventions, or considering ways to remove the burden of behavior from the consumer altogether (Pickering *et al.* 2019). For example, creating operational efficiencies, offering subsidies, and increasing regulatory enforcement may make adequate fecal sludge management possible in some settings (e.g., Burt *et al.* 2019).

There were several limitations in the current study. The most important is that while we can draw conclusions about WTP estimates from different preference elicitation methods, interpreting what true (often called 'normative') preferences are and reasons for estimates differing (e.g., due to social desirability bias) is subjective. Beyond showing where different methods agree and disagree, direct testing of these hypotheses would produce better estimates of true WTP – for example, by randomly allocating some plots to receive a particular upgrade and looking at rental prices achieved over time. Additionally, eliciting preferences via CVM and/or DCEs for attributes of rentals that were better estimated by HPM (e.g., number of rooms) or capturing the overall rental value through stated preference methods would allow for better comparison with real-world data, as HPM estimates were poorly specified in some cases for reasons described above. Second, each of these methods is subject to assumptions made by functional forms assumed and estimation methods used. Several models were estimated for each method, but different ways of constructing choice experiments, phrasing CVM questions, or categorizing plot characteristics for HPM may have led to different conclusions. However, we used standard approaches for each of these three methods to generate results that were as policy relevant as possible.

For policymakers, it is clear that CVM alone is inadequate to determine demand for services. While revealed preference methods are not feasible in many settings where new products or services are to be provided, DCEs and other less-direct methods should be considered. As was shown above the problem is not simply that CVM estimates may be inflated, leading policymakers to rely on them to indicate perhaps only relative preferences between options, but instead may produce varying levels of bias in estimates or even estimates that are of the wrong sign. Since DCEs allow for the estimation of tradeoffs between attributes, and hedonic pricing ensures that the overall rental fees paid are consistent with market prices, combining the results of these approaches may be the best way to accurately estimated WTP, with CVM being reserved for testing new or hypothetical products or services where presenting choices between alternative services is not a straightforward exercise. Future research should further validate this approach using DCEs where more clearly defined housing attributes (e.g., rent paid per room of housing) can be compared to results from hedonic pricing.

Despite these methodological challenges, it is clear that substantial willingness to pay for sanitation quality exists in peri-urban shared sanitation settings, and future research should build on this work by using incentive-compatible methods to firmly establish the potential market for sanitation in these marginalized communities and further validate indirect stated preference elicitation methods for estimating demand for new and existing water and sanitation products and services at low cost to those conducting the assessments.

CONCLUSION

Tenants expressed significant willingness to pay for several important aspects of higher-quality shared sanitation. However, some limitations were demonstrated with each method of willingness to pay assessment, and especially with the inconsistent bias displayed by contingent valuation. While the magnitude of these amounts will not be sufficient to reach high-quality sanitation for all, they may make a major contribution to gaps in sanitation financing when combined with demand-side approaches, strengthened markets, and subsidies. Accurate assessments of willingness to pay, likely using combinations of DCEs and appropriate revealed preference methods, will be a key consideration for peri-urban sanitation planning to reach the goal of sustainable sanitation services for all.

SUPPLEMENTARY MATERIAL

The Supplementary Material for this paper is available online at https://dx.doi.org/10.2166/washdev.2020.072.

REFERENCES

Acey, C., Kisiangani, J., Ronoh, P., Delaire, C., Makena, E., Norman, G., Levine, D., Khush, R. & Peletz, R. 2019 Cross-subsidies for improved sanitation in low income settlements: assessing the willingness to pay of water utility customers in Kenyan cities. *World Development* **115**, 160–177.

Altaf, M. A. & Hughes, J. A. 1994 Measuring the demand for improved urban sanitation services: results of a contingent valuation study in Ouagadougou, Burkina Faso. *Urban Studies* **31** (10), 1763–1776. doi:10.1080/00420989420081621.

Andrews, R. L. & Currim, I. S. 2003 A comparison of segment retention criteria for finite mixture logit models. *Journal of Marketing Research* **40** (2), 235–243.

Antwi-Agyei, P., Monney, I., Dwumfour-Asare, B. & Cavill, S. 2019 Toilets for tenants: a cooperative approach to sanitation bye-law enforcement in Ga West, Accra. *Environment and Urbanization* **31** (1), 293–308.

Ben Yishay, A., Fraker, A., Guiteras, R., Palloni, G., Shah, N. B., Shirrell, S. & Wang, P. 2017 Microcredit and willingness to pay for environmental quality: evidence from a randomized-controlled trial of finance for sanitation in rural Cambodia. *Journal of Environmental Economics and Management* **86**, 121–140. doi:10.1016/j.jeem.2016.11.004.

Beshears, J., Choi, J. J., Laibson, D. & Madrian, B. C. 2008 How are preferences revealed? *Journal of Public Economics* **92** (8), 1787–1794.

Boyle, K. J. 2017 Contingent valuation in practice. In: *A Primer on Nonmarket Valuation* (P. A. Champ, K. J. Boyle & T. C. Brown, eds.) Springer Nature, The Netherlands, pp. 83–131.

Brueckner, J. K. 2013 Slums in developing countries: new evidence for Indonesia. *Journal of Housing Economics* **22** (4), 278–290.

Burt, Z., Sklar, R. & Murray, A. 2019 Costs and willingness to pay for pit latrine emptying services in Kigali, Rwanda. *International Journal of Environmental Research and Public Health* **16** (23). doi:10.3390/ijerph16234738.

Cameron, T. A., Poe, G. L., Ethier, R. G. & Schulze, W. D. 2002 Alternative non-market value-elicitation methods: are the underlying preferences the same? *Journal of Environmental Economics and Management* **44** (3), 391–425.

Carlsson, F. & Martinsson, P. 2001 Do hypothetical and actual marginal willingness to pay differ in choice experiments: application to the valuation of the environment. *Journal of Environmental Economics and Management* **41** (2), 179–192.

Carson, R. T., Flores, N. E. & Meade, N. F. 2001 Contingent valuation: controversies and evidence. *Environmental and Resource Economics* **19** (2), 173–210. doi:10.1023/A:1011128332243.

Champ, P. A., Boyle, K. J. & Brown, T. C. 2017 *A Primer on Nonmarket Valuation*. 2nd edn.

Chopra, V. & Das, S. 2019 *Estimating Willingness to Pay for Wastewater Treatment in New Delhi: Contingent Valuation Approach.*

Cohen, J. L. 2019 The enduring debate over cost sharing for essential public health tools. *JAMA Network Open* **2** (8), e199810.

Danert, K. & Hutton, G. 2020 Shining the spotlight on household investments for water, sanitation and hygiene (WASH): let us talk about HI and the three 'T's. *Journal of Water, Sanitation and Hygiene for Development.* doi:10.2166/washdev.2020.139.

de Bekker-Grob, E. W., Donkers, B., Bliemer, M. C. J., Veldwijk, J. & Swait, J. D. 2020 Can healthcare choice be predicted using stated preference data? *Social Science & Medicine* **246**, 112736. doi:10.1016/j.socscimed.2019.112736.

Evans, B., Hueso, A., Johnston, R., Norman, G., Pérez, E., Slaymaker, T & Trémolet, S. 2017 Limited services? The role of shared sanitation in the 2030 agenda for sustainable development *Journal of Water Sanitation and Hygiene for Development* **7** (3), 349–351. doi:10.2166/washdev.2017.023.

Ezeh, A., Oyebode, O., Satterthwaite, D., Chen, Y.-F., Ndugwa, R., Sartori, J., Mberu, B., Melendez-Torres, G., Haregu, T. & Watson, S. I. 2017 The history, geography, and sociology of slums and the health problems of people who live in slums. *The Lancet* **389** (10068), 547–558.

GBD Diarrhoeal Diseases Collaborators 2017 Estimates of global, regional, and national morbidity, mortality, and aetiologies of diarrhoeal diseases: a systematic analysis for the Global Burden of Disease Study 2015. *The Lancet Infectious Diseases.*

Gulyani, S., Bassett, E. M. & Talukdar, D. 2012 Living conditions, rents, and their determinants in the slums of Nairobi and Dakar. *Land Economics* **88** (2), 251–274.

Hanemann, W. M. 1994 Valuing the environment through contingent valuation. *The Journal of Economic Perspectives* **8** (4), 19–43.

Hanemann, M. & Kanninen, B. 2001 11 The Statistical Analysis of Discrete-Response CV Data. In: *Valuing Environmental Preferences: Theory and Practice of the Contingent Valuation Method in the US, EU, and Developing Countries* (J. Bateman & K. G. Willis, eds). Oxford University Press, Oxford. p. 302.

Hanna, N. & Richards, D. 2014 The impact of communication on a human-agent shared mental model and team performance. In *Paper Presented at the Proceedings of the 2014 International Conference on Autonomous Agents and Multi-Agent Systems.*

Harrison, D. & Rubinfeld, D. L. 1978 The distribution of benefits from improvements in urban air quality. *Journal of Environmental Economics and Management* **5** (4), 313–332.

Hartung, C., Lerer, A., Anokwa, Y., Tseng, C., Brunette, W. & Borriello, G. 2010 Open data kit: tools to build information services for developing regions. In *Paper Presented at the Proceedings of the 4th ACM/IEEE International Conference on Information and Communication Technologies and Development.*

Hausman, J. 2012 Contingent valuation: from dubious to hopeless. *Journal of Economic Perspectives* **26** (4), 43–56.

Hensher, D. A. 2010 Hypothetical bias, choice experiments and willingness to pay. *Transportation Research Part B: Methodological* **44/** (6), 735–752.

Hole, A. R. 2007 A comparison of approaches to estimating confidence intervals for willingness to pay measures. *Health Economics* **16** (8), 827–840.

Horiuchi, Y., Markovich, Z. & Yamamoto, T. 2018 *Does Conjoint Analysis Mitigate Social Desirability Bias?*

Hutton, G. & Varughese, M. 2016 *The Costs of Meeting the 2030 Sustainable Development Goal Targets on Drinking Water, Sanitation, and Hygiene.* World Bank, Water and Sanitation Program, Washington, DC.

Isacsson, G. 2007 The trade off between time and money: is there a difference between real and hypothetical choices? In: Statens väg-och transportforskningsinstitut.

Johansson-Stenman, O. & Svedsäter, H. 2003 *Self Image and Choice Experiments: Hypothetical and Actual Willingness to Pay.* Department of Economics, School of Economics and Commercial Law, Göteborg University, Gothenburg.

Knight, J. R., Herrin, W. E. & Balihuta, A. M. 2004 Housing prices and maturing real estate markets: evidence from Uganda. *The Journal of Real Estate Finance and Economics* **28** (1), 5–18.

Krinsky, I. & Robb, A. L 1986 On approximating the statistical properties of elasticities. *The Review of Economics and Statistics* **68** (4), 715–719.

List, J. A. & Gallet, C. A 2001 What experimental protocol influence disparities between actual and hypothetical stated values? *Environmental and Resource Economics* **20** (3), 241–254.

Little, J. & Berrens, R. 2004 Explaining disparities between actual and hypothetical stated values: further investigation using meta-analysis. *Economics Bulletin* **3** (6), 1–13.

Lixil, WaterAid, and Oxford Economics 2016 The true cost of poor sanitaion. Accessed on 23 September 2020 at: http://www.publicfinanceforwash.com/sites/default/files/uploads/LIXIL-WaterAid-2016-true%20economic%20cost%20poor%20sanitation.pdf

Mitchell, R. C. & Carson, R. T. 1989 *Using Surveys to Value Public Goods: The Contingent Valuation Method.* Resources for the Future, Washington, DC.

Murphy, J. M., Pagano, M. E., Nachmani, J., Sperling, P., Kane, S. & Kleinman, R. E. 1998 The relationship of school breakfast to psychosocial and academic functioning: cross-sectional and longitudinal observations in an inner-city school sample. *Archives of Pediatrics & Adolescent Medicine* **152** (9), 899–907.

Murphy, J. J., Allen, P. G., Stevens, T. H. & Weatherhead, D. 2005 A meta-analysis of hypothetical bias in stated preference valuation. *Environmental and Resource Economics* **30** (3), 313–325.

Nakagiri, A., Niwagaba, C. B., Nyenje, P. M., Kulabako, R. N., Tumuhairwe, J. B. & Kansiime, F. 2016 Are pit latrines in urban areas of Sub-Saharan Africa performing? A review of usage, filling, insects and odour nuisances. *BMC Public Health* **16** (1), 120. doi:10.1186/s12889-016-2772-z.

Null, C., Kremer, M., Miguel, E., Hombrados, J. G., Meeks, R. & Zwane, A. P. 2012 *Willingness to Pay for Cleaner Water in Less Developed Countries: Systematic Review of Experimental Evidence*. The International Initiative for Impact Evaluation (3iE).

Peletz, R., Cock-Esteb, A., Ysenburg, D., Haji, S., Khush, R. & Dupas, P. 2017 Supply and demand for improved sanitation: results from randomized pricing experiments in rural Tanzania. *Environmental Science & Technology* **51** (12), 7138–7147.

Pickering, A. J., Crider, Y., Sultana, S., Swarthout, J., Goddard, F. G. B., Anjerul Islam, S., Sen, S., Ayyagari, R. & Luby, S. P. 2019 Effect of in-line drinking water chlorination at the point of collection on child diarrhoea in urban Bangladesh: a double-blind, cluster-randomised controlled trial. *The Lancet Global Health* **7** (9), e1247–e1256. doi:10.1016/S2214-109X(19)30315-8.

Quaife, M., Terris-Prestholt, F., Di Tanna, G. L. & Vickerman, P. 2018 How well do discrete choice experiments predict health choices? A systematic review and meta-analysis of external validity. *The European Journal of Health Economics* **19** (8), 1053–1066.

R Core Team. 2018 *A Language and Environment for Statistical Computing*. R Foundation for Statistical Computing, Vienna, Austria.

Rosen, S. 1974 Hedonic prices and implicit markets: product differentiation in pure competition. *Journal of Political Economy* **82** (1), 34–55.

Schmidt, J. & Bijmolt, T. H. 2019 Accurately measuring willingness to pay for consumer goods: a meta-analysis of the hypothetical bias. *Journal of the Academy of Marketing Science* **48** (3), 499–518.

Scott, P., Cotton, A. & Sohail Khan, M. 2013 Tenure security and household investment decisions for urban sanitation: the case of Dakar, Senegal. *Habitat International* **40**, 58–64. doi:10.1016/j.habitatint.2013.02.004.

Sehreen, F., Masud, M. M., Akhtar, R. & Masum, M. R. A. 2019 A contingent valuation approach to evaluating willingness to pay for an improved water pollution management system in Dhaka City, Bangladesh. *Environmental Monitoring and Assessment* **191** (7), 457.

Shiras, T., Cumming, O., Brown, J., Muneme, B., Nala, R. & Dreibelbis, R. 2018 Shared latrines in Maputo, Mozambique: exploring emotional well-being and psychosocial stress. *BMC International Health and Human Rights* **18** (1), 30.

Simiyu, S., Swilling, M., Rheingans, R. & Cairncross, S. 2017 Estimating the cost and payment for sanitation in the informal settlements of Kisumu, Kenya: a cross sectional study. *International Journal of Environmental Research and Public Health* **14** (1). doi:10.3390/ijerph14010049.

Tidwell, J. B., Chipungu, J., Chilengi, R. & Aunger, R. 2018a Assessing peri-urban sanitation quality using a theoretically derived composite measure in Lusaka, Zambia. *Journal of Water, Sanitation and Hygiene for Development* **8** (4), 668–678. doi:10.2166/washdev.2018.029.

Tidwell, J. B., Chipungu, J., Chilengi, R., Curtis, V. & Aunger, R. 2018b Theory-driven formative research on on-site, shared sanitation quality improvement among landlords and tenants in peri-urban Lusaka, Zambia. *International Journal of Environmental Health Research* **29** (3), 312–325. doi:10.1080/09603123.2018.1543798.

Tidwell, J. B., Chipungu, J., Bosomprah, S., Aunger, R., Curtis, V. & Chilengi, R. 2019a Effect of a behaviour change intervention on the quality of peri-urban sanitation in Lusaka, Zambia: a randomised controlled trial. *The Lancet Planetary Health* **3** (4), e187–e196. doi:10.1016/S2542-5196(19)30036-1.

Tidwell, J. B., Terris-Prestholt, F., Quaife, M. & Aunger, R. 2019b Understanding demand for higher quality sanitation in peri-urban Lusaka, Zambia through stated and revealed preference analysis. *Social Science & Medicine* **232**, 139–147. doi:10.1016/j.socscimed.2019.04.046.

UN-HABITAT. 2003 *The Challenge of Slums: Global Report on Human Settlements*. Retrieved from http://reliefweb.int/sites/reliefweb.int/files/resources/7E2BED9A93F54E53432576030048C121-UNHabitat_Dec2003.pdf

van den Berg, C. & Nauges, C. 2012 The willingness to pay for access to piped water: a hedonic analysis of house prices in southwest Sri Lanka. *Letters in Spatial and Resource Sciences* **5** (3), 151–166. doi:10.1007/s12076-012-0079-y.

Van Minh, H., Hung, N. V., Thanh, N. H. & Yang, J. C. 2013 Assessing willingness to pay for improved sanitation in rural Vietnam. *Environmental Health and Preventive Medicine* **18** (4), 275–284. doi:10.1007/s12199-012-0317-3.

Vásquez, W. F. & Beaudin, L. 2020 A hedonic valuation of sanitation services in Guatemala. *Environment and Development Economics* **25** (4), 378–398.

Whittington, D., Lauria, D. T., Wright, A. M., Choe, K., Hughes, J. A. & Swarna, V. 1993 Household demand for improved sanitation services in Kumasi, Ghana: a contingent valuation study. *Water Resources Research* **29** (6), 1539–1560. doi:10.1029/93WR00184.

First received 26 March 2020; accepted in revised form 10 August 2020. Available online 19 September 2020

IWA Publishing's authorised EU representative for General Product Safety Regulations is Diane D'Arras, 15 rue Duret, 75116 Paris, France, e-mail: safety@iwap.co.uk.

Printed and bound by CPI Group (UK) Ltd, Croydon, CR0 4YY
12/05/2026
02108886-0004